The Smartphone Medic

The Smartphone Medic

Smartphones in ski resorts – save lives, reduce costs and boost your reputation

Duncan Isaksen-Loxton

Rasmus Pty Ltd 2014

Contents

Introduction

To start this book we need to understand a little about the scale of the industry and the impact we may be able to have. We also need to understand a little about the problems we face as an industry to be able to develop a strategy to improve and expand regardless of other external factors like the economy.

From my seat there is one clear path we can take to give us more certainty, and that is through the safety of our guests. It is key to getting new, younger customers interested in the sport (through peace of mind for parents), and keeping them coming back year after year.

It is unreasonable for me to believe that we can reduce accidents to close to zero because of the factors at play, but I do believe we can make significant improvement to resort business and improve the industry.

The Smartphone Medic will take you step by step through the issues confronting the industry with regard to risk management, data collection, training and education, and technology and present information on the latest app, Medic52, designed to overcome many of these issues.

You will discover how collected data can be used to promote your brand and position your business in front of your competitors; reduce your workload; manage and understand risk; educate staff and customers to help you and reduce insurance premiums.

Most importantly, you will see how implementing these steps will result in fewer accidents, create more loyal repeat customers, and make our skiing lives safer.

This book draws on the knowledge I have gained as a ski instructor, patroller and from my professional career as a business owner and software developer. In the following pages I aim to combine what I have learned – the practical and the theory – to offer solutions you can implement with ease.

Overview of the Snow Sports Industry

It is important to gain a global view of the ski industry - where the resorts are, who owns what, what regions are experiencing growth, what regions are experiencing contraction, ski visits and income generated by the industry. Where possible I have included accident statistics.

Research on this topic is a significant challenge, as lots of information is gathered in different ways by different organisations and countries. Many resorts don't like to share statistics on visits or accidents because of confidentiality issues. Much of the data supplied to industry influencers is voluntary, possibly inflated, or significantly undervalued.

How many snow goers are there?

Some estimate that there are 378 million skier days worldwide[1] but given the information is voluntarily supplied and with no way to cross check, there is no way to verify that figure for accuracy.

The majority of the market is focussed on France, Austria,

Switzerland and the USA. These four nations have 26% of the world's resorts, and over half (52.62%) of the reported skier visits. China is a rapidly growing market with a growth of 5 million skier visits to 7 million from 2008 to 2013.

The majority of customers are domestic to the resort, with only Austria, Andorra and Switzerland pulling more than 50% skier visits from foreigners. This is due to the small population (Andorra) and the significant popularity of these locations with the two largest foreign ski tourists from Great Britain and Germany.

The most prolific skiers by population are the Austrians, Norwegians and Swiss, but the British do remarkably well (given the distances involved) with 49% of the 1.2 million market claiming to have done more than 20 weeks of skiing.[2]

The problems start to show when you look at the number of days per season. In China the culture is such that people will go for one day, in their lifetime, to say they have done it. It is not seen as a lifestyle or ongoing sport. In the USA the National Ski Areas Association shows that 54% of skier visits are one day in a season (for household income greater than $100k), and that the best chance of creating a return visitor is 5 to 7 days per season.[3]

The 2013 Ski Club of Great Britain consumer survey showed a worrying 26% of respondents do not intend to go skiing again. The reasons for this are unclear.

Accident rates

Very little is known or published about accident rates across the industry. Most evidence is anecdotal, but from my research I mostly see numbers between 2.5 – 3 per 1,000 skier visits globally.

I can understand why these figures are not published. Resorts are reluctant to tell potential customers the number of people that get hurt. Imagine my wife's reaction when, as a new patroller, I came home and told her about all the things I saw that day. She didn't want to ski and marked it as a dangerous sport. With perspective though skiing is a relatively safe sport when compared to other outdoor pursuits.[4]

In the USA for example, 2010-11 saw about 11 million skier days, and 16,948 (0.15%) ended up in hospital with head injuries. By comparison, about 18 million played football – and 46,948 (0.26%) sustained head injuries. For cycling, the figures were 46.8 million and 85,389 (0.18%) respectively.[5]

Europe

Medecins de Montagne, a group of French Alpine doctors (*www. mdem.org*) show that the average skier can only expect to be injured around three times for every 1,000 days on the slope, while death is even less probable. In the Austrian Alps, 38 people were killed in 2012/13, compared to around 20 each year in France. However the proportions of serious injuries in France has climbed over the past twelve years, with those "heavily wounded" representing 5.2% of all injured skiers last year, compared to 3.95% in 2001, meaning 1,000 more skiers were hospitalised as the result of an accident last year than 12 years ago.

The Ski Club of Great Britain data shows that although skiing and snowboarding are active adrenaline sports, only two or three injuries occur per 1,000 ski days.

USA

The overall risk of injury from skiing remains unchanged in recent years at around 2.5 per 1000 skiers. In simple terms this means that for every 1,000 skiers on the mountain two or three of them will require some form of medical treatment.

The chance of dying in a ski accident in the USA remains very low. Over the past decade the *National Ski Areas Association* (NSAA) count has averaged 39.6 annual snowboard and skiing deaths in the USA, or 0.69 deaths per million skier visits. Even with steadily increasing helmet use (70% of all participants, nearly triple the number from 2003), the number has jumped around. There were 25 deaths in the 2012-13 season; 46 the season before; as few as 22 seven years ago.[6]

According to the US Consumer Product Safety Commission, there were more than 144,000 skiing related injuries treated in hospitals, doctors' offices, and emergency rooms in 2010. Injuries associated with snowboarding were even greater at 148,000.

Falls are an obvious cause of injuries, accounting for approximately 75% to 85% of skiing injuries. Collisions with objects, including other skiers, account for between 11 and 20%, while incidents involving ski lifts contribute between 2 and 9%. The majority of injuries are sprains, followed by fractures, lacerations and dislocations.

Canada

In 2010–2011, 2,329 people were admitted to hospital for a skiing or snowboarding fall or crash from the 19 million skier visits reported. 415 were hospitalised for head injuries related to a winter sport or recreational activity (including hockey, snowmobiling & skating). Almost a third (135) of these serious head injuries happened to skiers and snowboarders. Over the past five years, a total of 759 head injury hospitalisations happened on the hill.

In contrast to findings in the USA, helmets have been linked to a 35% reduction in head injury risk for skiers and snowboarders. [7]

Ski Areas – challenges and industry operation

There are approximately 6000 ski areas worldwide, across 49 countries, with only 2000 having more than five lifts. 50% of the world's resorts are in Western Europe.[8]

Resorts can be found in some surprising locations: in Australia you can ski among gum trees and need to look out for wombats, and in India and Iraq you will see armed soldiers on skis.

Sunny Spain has 34 resorts, and there is a 380 metre World Cup slalom run on the island of Cyprus. Greece is associated with beaches and islands, but mountains cover 80% of the country. China contains nearly 600 resorts with only 11 having over 300m of vertical. Australia is the home to Michael Milton, the fastest disabled downhill skier at 213.65km/h.

Indoor skiing is popular too. The first centre opened in Japan in 1959, and you can now ski indoors in Dubai. The UK has about 60 outdoor 'dry' slopes, and six indoor snow centres. Lithuania has an indoor slope that opened in 2011, and is one of few in the world to also be open air in the winter.

The most advanced infrastructure (lifts, snow making etc) is in Austria, with 70% of terrain covered with snow making (with 700 million Euro/USD $946M of investment since 2008) and the most up-to-date lifts. This is perhaps not surprising as the native lift manufacturer, and industry leader, Dopplemayr, was founded in Austria in 1892.

The most successful resorts have dealt with a lack of snow by branching out into a year round operation, and developing their revenue streams.

Generally, resorts rely on three revenue streams:
- Residents
- Visitor revenue
- Real estate

In summer a range of activities bring in visitors, such as:
- golf
- mountain biking
- zip lines
- summer luge
- bush walking and hiking

Winter sees a change in activities to:
- skiing
- snowboarding
- tubing
- cross country skiing
- snowshoeing
- snowmobiling

According to international consultant, Laurent Vanat, "Nearly everywhere, the industry is facing the challenge of generating long term growth."[9]

The industry is suffering from Baby Boomer syndrome. Baby Boomers make up the current majority of enthusiastic participants. This generation will start to exit the market soon without adequate replacement from younger customers. The need to generate enthusiasm in the younger set is extremely important and critical to the survival of the industry.

Vanat also points out that this is far from a done deal. Many resorts that have been trying for years to convert new customers into loyal participants have some way to go. "It requires a significant effort to create a situation that only improves very slowly."[10]

To illustrate that with a social media statistic, the world's most popular and well-known ski resorts are barely able to attract 150,000 friends to their Facebook pages, even though they host more than two million skier visits per year.

Resorts – who owns what?

Skiing has grown up in many different ways around the world, some organically from mountain villages, through to purpose built ski resorts like Sochi in Russia, host of the 2014 Winter Olympics.

Europe
France has the world's largest ski area operator in the world – Compagnie des Alpes. They operate 14 of the significant French ski resorts such as La Plagne , Val d'Isere and Chamonix.

Switzerland, Italy, and Austria have very fragmented ownership models, having started as mountain villages and farmland, growing up to be ski destinations. These locations prove to be difficult to manage with many stakeholders involved such as landowners, farmers and tourist operators. Austria seems to have it well managed given their current standing in the world market by visitor numbers.

Germany has more than 500 resorts, but about 50% of these only have one lift. There are two or three significant resorts in the south including Garmisch Partenkirchen that hold a regular world cup race each year.

Norway has two resorts with a unique snow guarantee: Trysil and

Hemsedal guarantee that if ski trails are not open for an extended period of time, the resort refunds their guests the money for their hotel, ski school, ski rentals and the lift pass.

Spain has one of the top 50 largest resorts, Sierra Nevada, which is also the southernmost in Europe, in the second highest mountain chain in Europe. It's only an hour drive from the Mediterranean making it one of the few places you can surf and ski in a day.

The Lauberhorn in Wengen, Switzerland is home to the longest downhill world cup race.

USA

Many resorts are small, and independent, with a few exceptions where they are slowly being bought up by corporations such as CNL Lifestyle properties (24 resorts including Crested Butte), Boyne (10 resorts including Big Sky, Montana), Powder Corp (nine resorts including Copper Mountain) and Vail Resorts (10 resorts including Beaver Creek). These companies deal in more than ski resorts with diversification into other types of leisure operations and hotels/resorts.

Squaw Valley is home to the world's only ski in/ski out Starbucks.

Canada

The industry is mostly split geographically, east and west, with the majority of resorts owned by Mont Saint-Sauveur International or Resorts of the Canadian Rockies LTD.[11]

China

As a developing nation of nearly everything it's no surprise to find China has 523 resorts, but only 60 have one or more lifts, and only 11 have 300m of vertical drop or more. They are however turning over good numbers with 16 million skier days last winter with 20% return rate domestically.[12]

China has the highest gondola in the world – Dagu Glacier Ropeway, Sichuan province, with a summit at 4,843 metres.

Japan

Japan has had some economic issues and natural disasters, and this has directly impacted the ski economy. Eighty resorts were closed during 1996-2008 and there are now 821 resorts in total.[13]

South Korea

South Korea has a well developed ski economy opening its first resort in 1975. In one resort they have night skiing until 4am, 70+% of the market is snowboarders and young people, driven in by a big schools program to drive passion in new skiers.[14]

South Africa & Lesotho

The South African resort of Tiffindell runs the only FIS races in South Africa. Afriski, in nearby Lesotho is at 3,000m and six hours drive from Johannesburg.

South America

South America is home to the most southern most resort on the planet, Cerro Castor, and also the newest resort in Argentina. The first lift was built in 1939 at Catedral, near Bariloche which is now considered to be the Chamonix of South America.

Valle Nevado is 3,830 metres above sea level, and Portillo in Chile has a maximum capacity of 480 guests in the resort across two lodges and the famous yellow Hotel Portillo.

Staff employed by the skiing industry

There are many different jobs that make the ski industry work, inside and outside the resort. On snow we have groomers, snow makers, patrollers, instructors, lift attendants and maintenance crews. Off snow there is hospitality, sales, equipment and rentals. Beyond the resort there are travel agents, insurers, marketing, equipment, airlines and bus companies. These are just a few of the people that help get a person to become a customer and get to the top of the mountain for their ride down.

Many of these staff work seasonally, following the snow from north to south, and working in a job that is their passion.

An idea of the importance of the snow industry to a community is in Alberta, Canada. In a report from 2000 that states that approximately 12,300 person years of employment Canada-wide were sustained by the expenditures of skiing at Alberta Rocky Mountain Ski Resorts.[15] Many of these staff would be seasonal, but it illustrates how many people are involved.

CASE STUDY: ROSSIGNOL

Rossignol is a well-known French ski and board manufacturer, started in 1907 by Abel Rossignol, a carpenter who crafted his first pair of wooden skis in 1937.

After a brief period of ownership by surf brand Quicksilver, it is now owned in part by the Boix-Vive family and an investment group that owns Helly Hansen clothing.

Rossignol owns a number of significant brands in the industry – Dynastar skis, Look bindings, Lange boots, making it a powerhouse of equipment manufacturing.

Quicksilver took manufacturing out of France, and the new owners have invested nearly $13 million in factory upgrades through 2012 to bring manufacturing back to France and create jobs.[16] They now manufacture in France and Spain with about 1,500 people around the globe, of which half are in France.

Rossignol manufactures nearly one million skis a year and, in 2012, posted a profit of €5 million.[17]

This company has a considerable reliance on ski resorts continuing to grow and operate.

Income generated by snow sports

The industry represents significant revenue throughout the world.

In the USA, snow speciality retail stores reported USD $1,783 million ($1.8 billion) in sales in 2012.[18]

The big resort groups post significant numbers too. Vail Resorts posted USD $37.7 million profit in 2013 with revenue of $1.12 billion from its nine resorts.[19]

The US ski resorts in 2011 reported revenue of $2.6 billion, with net profit estimated at $182.6 million.[20]

This is remarkable given that an estimated 50% of revenue for a resort comes from lift ticket sales[21] and that 85-100% of income is generated in around 140 days per year depending on snow season length.

Costs of Accidents to the Industry

Ski resorts must find ways to reduce the number of accidents. The cost of an injury to the resort is significant in cash terms if the resort is held liable in a negligence claim. This will result in an increase in insurance premiums, and possible loss of the customer and their extended family.

However, if the customer has a great experience and becomes a lifelong safe skier or snowboarder, the benefit to a resort, and the industry as a whole, is huge.

For your resort, the greatest loss incurred is the impact on the number of return visits or the drop in frequency of returns. If a guest is seriously injured they could be out of action for a week, the season, a few years or forever. For a family unit, an injury may affect the entire family's ability to go skiing again. They may perhaps choose to go to another resort based on the events of that day.

So where is the biggest cost? Is it in the potential damage to income through the loss of a lifelong skier? Is it a loss to the overall industry when a skier is unable to continue? Is it the cost associated in taking the matter to court and the liability payout that you may incur? Is it the cost of increasing insurance premiums?

Resort marketing company Ryan Solutions[22] noted in 2013 that if a guest has come to you for two years in a row there is a 34% chance they will return next year. If they do so for six years, the chances of them coming back to you shoot up to 75%. What is one guest worth to your resort? What if that guest is connected to a family of four and one accident stops the whole family from creating a tradition with you?

Consider the loss across the wider industry, like travel, accommodation and equipment industries. These businesses all help a resort to get the customer to the top of the hill. A guest who skis once or twice a year at a resort in their home country, and takes another trip to a foreign country, will affect airlines, accommodation providers, restaurants and bars, entertainment, lift ticket sales, rentals, insurance purchases, equipment manufacturers, lessons and training. The list goes on. The footprint of that loss to our industry is substantial.

Costs to the resort

When a ski resort relies on a significant amount of its income to come from lift ticket sales, having people stay on the slopes, and return is a business priority.

In US resorts, lift ticket sales made up just over 50% of income[23] with food and beverage and instruction another 21% combined.

With the average weekly ticket costing $580, having a family of four not return for their next annual ski trip (or two), the resort stands to loose $4,640 in ticket revenue over the next two years. If we also assume that those two years were critical for the two children taking up the sport for life, then in today's dollars this one event cost nearly $35,000 over 30 years.

If resorts can get some of the many casual one day customers to get a few days under their belt then there is a greater chance of them becoming hooked, and returning. Ryan Solutions note that the sweet spot for renewals is people who ski between five and seven days[24] so resorts must start to understand why the single day skiers are not doing more, and work to convert them to multi day visitors who come back year after year. This is only possible if we work to keep them safe whilst in our care.

Costs to the customer

An accident can impact a patient's life in different ways depending on the nature of the injury. For someone who works in a manual trade, or relies on their fitness like a sports person, any injury will affect their ability to earn and support their lifestyle and family. This can also have a knock-on effect on their family who may have to support the injured person through their time of recovery, or for life in the case of a totally disabling injury.

An accident on holiday is also going to have a significant effect on the customer. One of my fellow Thredbo patrollers, John Hazell, had an accident on his first day skiing while on a trip to New Zealand in 2011.

"I was on the skier cross-course with my mates, and we came up over the jump, blind to the other side. A snowboarder stood in the middle of the trough. I was in the air and tried to avoid her, the best I could, but we collided and my leg and ski went between her legs. As we came down, her board and feet twisted around my leg and it shattered my ankle and tore my ACL."

As this was his first day, the rest of the trip was written off. Not only did John require care to get his leg strapped up and accumulate medical bills in a foreign country, but his ski pass investment was useless, and the rest of the holiday was spent finding entertainment in the local town waiting for friends to come back from a day's adventure.

At the time he thought it was only his knee that was injured. On returning home after visiting the specialists, he was diagnosed with a serious ankle injury.

This first day accident has culminated in a year of rehabilitation, physiotherapy and having to fund the gap between medical fees and his private medical insurance policy.

The injury has prevented John from working as a financial planner to his full capacity as he was not able to drive or walk unaided for three months. This has adversely impacted his income combined with lost time through the regular visits to a consultant, three major operations and regular physiotherapy.

The prognosis: Providing he has no complications from surgery and his body heals well, he should be back on the snow in 12 months.

John is a long-time skier and can think of nothing worse than having to stop skiing. Someone who is less competent and has had less investment in the sport may well have been turned off altogether at this point, and the industry would lose a customer for life.

The cost to John is significant when looking at all the factors.

ITEM	COST	TIME
Flights	$1,200	1 week
Accommodation	$600	1 week
Travel Insurance	$150	1 week
Lift passes	$500	1 week
Entertainment	$700	1 week
Rescue	Nil (under NZ accident scheme)	1 day
Initial treatment	Nil (under NZ accident scheme)	1 day
Australia Medical Insurance	$1,200	Annual
Surgery	$33,000	3 operations over the course of a year
Physiotherapy	$3,200	12 months
Loss of income	$30,000	6 months
Medical Insurance Coverage	+$18,500	
Total Cost:	**$50,050**	

It should also be noted that he was lucky to be in New Zealand where his rescue and initial treatment are covered by the New Zealand Accident Compensation Corporation. They cover travellers and residents to New Zealand and provide medical assistance when a person is involved in an accident. In return for getting free assistance, the individual's right to sue for personal injury is lost.

John's travel insurance was not used at all as with most travel insurance coverage it does not cover expenses when you return to your home country. All of the related expenses John incurred in Australia were out of his pocket or through his own standard medical insurance. Luckily he lives in a country where there is at least some government funding towards medical assistance. This cost would be much higher where no assistance is available.

The out of pocket expense to John was approximately $50,000 when we take into account loss of income. For many, this can be an annual wage, and would put the person in financial crisis, preventing them from making another ski trip for some time.

Thankfully, John has recovered well and was back on skis in the winter of 2013.

Insurance

When the insurance market is at an all-time high in terms of supply, the resort enjoys a good period of time where deals are easy to strike because there are lots of agents competing for the business. This provides a significant benefit to the insured resort because premiums are low and it's easy to get insurance. As we will see later, the industry goes in cycles and is affected by things such as world events and the stock market (to name only two). At the high side of the cycle, insured resorts don't have to worry too much about how to pay for the insurance. However, you should do something about how your insurance is structured today to protect against future premium rises caused by events outside of your control.

More accidents generally mean more payouts, and higher premiums. Insurers calculate their premiums by looking at the average annual claims cost, and the maximum amount of risk that could occur.

Researchers estimate that of the 66,000 injuries in Switzerland's resorts only 34,000 are actually insured.[25] The cost of the injuries tops $US 250 million to the insurance companies alone.

These injuries carry a significant risk of leading to litigation and negligence claims depending on local laws. Through sheer volume of skier days, the chances of an injury escalating to litigation is high. The cost of involving lawyers and the amount of time required by a team of employees to gather the evidence to defend a case is astounding.

Of course, by deciding to settle out of court, the company could end up with a lower cost. Even settling against a potentially large lawsuit still needs an investment of time and money in order to achieve that reduction. Keeping insurance is one way to provide peace of mind should this occur. This reduction however, is taken up by the insurance company if it is covered, but that cost will come back to the company at some point in the future through increased premiums.

In unforseen situations like natural disasters where the insurance industry pays out large returns, smaller companies often close down because they can't cover the claims. Premiums go up because the remaining companies can then dictate the terms. If you proactively manage your risks, data and systems, you will be able to prove to the insurer that they don't need to set their price to "cover the inevitable", and that incidents and accidents can be controlled or managed. Each and every accident adds to your potential claims cost, so working to keep each down means a better premium price upon renewal.

In Switzerland, certain alpine regions rely on the winter tourism as a vital part of their economy. In Bernese Oberland tourism contributes 26.6% to GDP [26] and thus the industry must be monitored and adjusted very carefully. Careless accidents and unforseen insurance premium increases could cripple this delicate economy.

Ways to be insured

1. Getting Insured through an Underwriter

There are two steps to getting insured. First you need some data to predict the loss that will occur over the period of insurance. A new ski resort would have no data, but could use data from the industry average, or a resort of a similar size, with similar demographics. If you've got no data, underwriters have to price for the worst. So they have to assume the gaps in the data could be negative. This results in a higher premium as the risk cannot be quantified.

> *"Underwriters will price for uncertainty if there's uncertainty there."* – Robin Barham (Lloyds Underwriter).[27]

Let's say that you are happy to cover losses up to 10% of your projected profits for the year. For example, a company looking at a $10 million profit may be able to look after losses up to $1,000,000. This would most likely be made up of many different sizes of claims across the period and not just one single large claim. To decide what this number looks like, I suggest you sit with your accountants to look at your books and work out what is possible.

You also have to balance that number with your projections for the coming year. If you have just had a bad snow year, your projection for the next year is probably uncertain and low. Your margins are going to be tight and therefore your appetite for risk is probably not all that high. Last year's profits won't allow you to use surplus cash to increase your excess (thus reduce your premium) either, because your shareholders are not getting back what they hoped for!

Once you have in mind how much you can cover yourself for, get in touch with a broker. You should maintain a few relationships with different brokers to work between them and understand what's going on in their world, and find one you can work with. Any major purchase like this should not be done with only one view. You should talk it through with a number of companies and as you do more, you will learn and understand more about their remit.

Another good reason to work with a few insurers is related to the market cycle. We may have another natural disaster and one of them could be out of business by the end of the year.

Take your time to understand where the insurance industry is in the cycle, how much supply is available and what the competitive environment is like. You can use this to understand what may occur in the next year and how it will affect your costs.

When presenting to your broker, make them understand what you do to manage your risks. Get them into the situation and show them how much you do to improve your guests' experience and make them safer. Things to consider are educational programs, data collection and preventative actions *based on data*.

2. Be a self insurer

One of the very cool things about insurance is that you can set up your own system to deal with it. This however requires that you have a ton of cash set aside to do nothing but be available should a catastrophe occur. What follows is simplified for the purposes of illustration and is not a detailed step by step on how to create your own system to self insure!

In order to work out the viability of being a self insurer, let's start by working out how much of your annual fees you could save if you lowered your premiums slightly. Let's say you took on slightly more risk in the next year by raising your excess limit with the insurance company you use and this saved $200,000 in insurance premiums. If you set the money aside and invest it in a safe vehicle, (not an armoured car) like a fixed interest cash deposit, which is guaranteed to make a few dollars each year, then you are on the road to being your own insurer.

If you also work on your systems and process around risk management and put in some really solid grounding to manage some of the issues in your resort, then you are lowering the chances of accident, and therefore the chances of needing to go to court or settle on a liability claim.

Combine the effects of these over a few years, and not only do you have a safer resort but your ability to pay less for your insurance is compounding on two fronts - financial and exposure to accidents.

This means as you go forward year on year you can reduce the amount you pay to the insurer on the basis of the practical work, and invest that saving into your own insurer. As you build that up, your ability to take on larger claims yourself improves.

Note that in our industry it is almost guaranteed that you will pay out that capital throughout the year. Your pace at which you grow your capital is dependant on reducing the number of claims you pay for, and the interest rate at which that capital can grow.

Be warned, self insurance is an inherently risky proposition, and one large claim could wipe it out. Thus there is a need to still cover that risk with another insurer, who in turn backs their risk with another. You will also need legal counsel on hand, as this would have been another factor taken care of by your insurer. Your legal advisor can advise on local laws and legislation in this area too.

How can you leverage your system to reduce your premiums

The less information you have, the less you can prove you are safe. In cases where there is only summary accident information available, the insurance company has little choice but to assume maximum risk and liability. They will assume performance based on industry trend and past experience or maybe not offer their support at all!

If you have been following along, then you will know that this means a larger sum of money from you to whoever does take on the risk.

Keep your vigilance

When we are in the cheapest time of the insurance cycle most people don't manage the risk well. They don't need to because there is no financial imperative to do so. Businesses focus on areas where larger cost savings can be made for the smallest amount of investment. Thus when premiums are low, investment can be directed elsewhere and the company can afford to be a little less vigilant because it's really very affordable.

These companies start to invest in staff and process when the premiums start to increase, and large settlements start to appear on the horizon. This is crazy if you think about it, because it increases costs by a factor of the premiums, legal fees, staffing and procedural change. Combined, this makes the entire situation much more stressful, increasing the chances that it could go wrong.

Managing the risk better, in an easier climate means you will have time to not only make change to improve, but also to collect data ready for the next renewal.

Making the change now may affect your bottom line a little, but not nearly as much as when you consider the alternative situation that can occur in the near future.

When the broker comes back to you, and they need to increase your premiums, you have the ability to present a case to not increase the amount, because you are being awesome!

Remember that the cost of insurance can go up because of events completely unrelated to you, and your situation. World changing events occur and affect the insurers, and this means premiums go up all round. So get on board and start to be prepared for the inevitable.

Time and resources

Ski patrollers are really effective when they are out on the hill. They help to keep the resort safe by monitoring changes in conditions and taking the appropriate action. When a rescue occurs and the patient is safe there is paper work to be done.

In 2011 the Mountain High ski patrol created an iPad app to help patrollers collect incident data and make it more accessible to mountain management and safety personnel.

Robert Chacon, Mountain High's assistant mountain manager at the time said in an interview with the NSAA Journal, March 2011:

"The integration of our electronic investigation database and iPad have decreased our time spent per incident by 75 percent. [...] What used to take us an hour for a complete incident and investigation now takes less than 20 minutes."

This means their patrollers are back on hill faster, and they have the reassurance that software is ensuring all the 'paper' work has been done to the standards required.

Resort reputation

There is the balance between attracting new visitors and keeping the existing ones through the use of marketing and word-of-mouth publicity. Some accidents will swing people away from a resort while other visitors will simply take it in their stride. Experiences that customers share with others can have a positive or negative effect. This is something that is now greatly magnified through social media.

Without the customer it is impossible for any business to exist. Therefore we can understand why we do not want publicity garnered from accidents. It's generally not a good thing. However there are ways to change this, and using best practice processes and procedures to stay in front of any negative publicity, and still produce a great experience for your guests. Most experienced customers know the risks, but they are not currently our issue in the industry. We need new customers from the non-skier population who are yet to understand the potential dangers of the sport. This is why it is essential to be prepared and manage press well.

Love them or hate them, there is no escaping the media. Thanks to multi-channel media outlets and round the clock coverage we are constantly in touch with 'world events'. How often do you think that there is more violence/natural disasters/cute kittens in the world than there used to be? It may be much the same, but technology now has the means to get it to us faster. In either scenario any hiccup or oversight can make the news, and once it is on the internet, it will never go away.

One reason could be perception. The news is consumed through many forms of media today, and it's accepted as fact much quicker. Media outlets look for sensationalist or high-impact news that pulls the heart strings and more often than not, is related to a company's poor performance or a human struggle.

Watching and reading the news is mostly a catastrophic event!

The public perception about our industry is critical to our survival. We cannot rely solely on the risk takers who hike the highest peaks, and get the adrenalin kick from a 60 degree, six-foot-wide chute surrounded by rocks. These people in our marketing and ski movies are the minority. However they set the standard and have the power to promote the right messages. Just as in any industry, our celebrities have leverage.

And just as a positive story can generate income, a negative story can have a huge impact on your bottom line.

A poorly managed accident can make the news in a matter of seconds – passers-by uploading footage to YouTube and to the media, conversations between rescuers and patients recorded and going viral, and the immediacy of a social media report appearing almost instantly on Facebook and Twitter.

Whether the reports are based in truth or not becomes irrelevant – the impact can have a ripple effect, damaging your brand and reducing visitors. In the 2013 Michael Schumacher accident, the investigators and resort got lucky because he was wearing a camera, and a bystander accidently filmed the Formula 1 star as the accident happened. This would definitely be an unusual occurrence no more than five years ago, but now the prevalence of mobile phones and sports cameras mean we could be filmed at any time.

Take a look at the resort responses and management to the recent accidents of high profile celebrities, like Michael Schumacher, Angela Merkhel and the deaths of Amanda Richardson in 2009 or Dutch Prince Friso in 2013. The resort and medical care providers are all under the scrutiny of the press and require almost instant information with great planning to manage the event. They must also go about their daily business and provide information from accident reports to investigators.

Workbook / Tips

- Write down what a visitor is worth to your resort.
- How many of the accidents you had last year prevented them from returning this year?

- What was the cost to your resort because of injury claims? How does this compare to previous years?
- Was there a common pattern to the injuries?
- What is your media management plan should a disaster or serious accident occur?
- How much time does a patroller, risk manager or accident investigator spend on an accident, and what can be done more efficiently?
- Are you in a position to consider self insuring?

Risk Management

Winter tourism is a big industry. Every year, about 115 million skiers seek out their next ski adventure. Beginner-grade skiers try out a new sport with no regulation and education, and novice participants using professional-grade equipment are increasingly putting themselves in dangerous situations through a lack of knowledge.

With snow users pushing beyond their limits every day, risk management is crucial to any on-snow operator, guide or safety professional.

The ski industry is unique to any other in the leisure sector in that you can almost guarantee a customer is going to get hurt every day the resort is open.

The leisure industry's primary focus is to create a safe environment that allows customers to push their boundaries but always come home after, and come back for more.

Consider skydiving, white-water rafting, abseiling, caving, and diving. All these activities require specialist equipment and knowledge before attempting them. Any provider of these exciting sports will take you through a safety briefing, talk to you about the risks and educate you before taking you out into the air/water/cave/foam. They take you through the process step-by-step to help you integrate the learning, to make sure you are safe and know how to follow the rules. For example, you cannot rent scuba gear unless you have a PADI or equivalent qualification. This is baked into the legislation – all in the name of safety.

However with skiing and snowboarding, everything is much simpler. There are no mandatory lessons, safety briefings, pre-training or age restrictions. You can walk into a rental shop, be fitted for skis and boots and you are free to have a go. You don't even need to use the facilities provided by a resort, because they are generally on public land or national parks. This combination of factors means the resort personnel have limited time to communicate with participants about the dangers and skills required to cope.

Skiing and snowboarding are dangerous sports. They can even be life threatening. There will always be accidents in ski resorts, but we *can* reduce risks for the benefit of the participants.

Managing risk, learning from accident events and providing safety for every snow user is a critical part of operating and working within the winter tourism industry.

Every snow operator – from resort owner to ski patroller to "liftie"– can take significant steps to improve their customers' experience on piste. This book provides some of the answers to these challenges and shows you how to use them to gain benefits in many other areas of the resort business.

Customers should come to your resorts knowing that they are in the safest possible place and are your highest priority. They should know they are not being put at risk or being manipulated out of money. You need to do everything you can to have them come back again and again, becoming life-long snow lovers. You may think that as an individual you are unable to influence how customers experience their time on the snow. However, every one of us can make a difference.

Most safety and risk management practices are formed in reaction to incidents, and legislation often resulting in defensive strategies. Resorts will do enough in risk management to meet the line that is set by local laws and legislation. Then they will move that line a little further by adding insurance. Once the line is well covered, that's where the work ends. By changing the focus and thinking about safety as an asset, a ski resort can multiply its benefits in areas such as cost, efficiency and marketing.

What is risk management?

Risk management is identifying the possibilities of accident or injury, and deciding what to do about them using factors such as proba- bility of occurrence, severity of injury and effort required to manage.

We use a risk management process to help define the boundaries of what is acceptable to a person or organisation. It is a way to reduce the complexities of a number of factors down into an easily under- standable system that any one can use to assess and manage risks.

Here is an example of a risk assessment process that captures a range of issues for consideration, extracted from Tracey Dickson & Tonia Grey – *Risk Management in the Outdoors*, 2011.

1. Establish context
Consider the function of your business. What is it you set out to achieve? When you think about this, what factors or elements will potentially expose you or your stakeholders to risk? Consider fac- tors in areas such as environment, social, legal, technology and your economy when looking for ideas.

Who are your stakeholders? Understand who may be affected, or who affects the risk or decisions made. This could be the guest who has an accident, or a staff member that makes the change leading to an accident. You may have volunteers such as mountain guides or volunteer patrollers who also need to be considered, as they will have different requirements with insurance and legislation.

Insurance companies, shareholders and the government may also be stakeholders in your situation.

This context helps you identify the people and places you need to observe, and the amount of risk you are geared up to manage.

2. Identify Risks

Risks that could arise from your activity can be in one of the following areas:

- **Ethical** – affecting reputation or belief.
- **Financial** – fraud, theft, membership fees, insurance, funding opportunities.
- **Physical** – personal injury or damage to physical assets.
- **Legal** – responsibility to the law, federal and state government.

You can find and identify risks by looking at data from past activity (see next chapter). Examine experiences and media coverage or interview the stakeholders. There are many ways to do this. Ask yourself how you can get the best information.

Tangible vs Intangible Risks

In your resort you may come across two types of risks.

Tangible risks can be defined prior to the event and have a fixed dollar or time value.

Tangible risks are quantifiable in cash or time. For example if a ski resort had no snow, we can predict how much revenue would be lost and the impact on the business. Another would be pricing pressure in the market, from overseas through globalisation or your competitor next door.

Consider the following types of tangible risk and how each may affect you:

- Relationship risk relates to tension between employees resulting in an unpleasant work environment and undermining of the workplace culture, leading to inefficiency in the business.
- Process-engagement risk occurs when the procedure carried out takes longer than necessary, or misses essential steps. This leads to lost time or incidents that could otherwise be avoided because the work was not completed properly.

- Knowledge risk refers to gaps in staff education resulting in staff dissatisfaction.

Intangible risks are outside of your control, such as a stock market crash where you cannot predict the cost.

Intangible risks have a 100% opportunity of occurring but the risk is ignored, due to lack of knowledge required to identify it. They cannot be reasonably quantified or predicted, and can include things such as regulatory changes, economic fluctuations and political volatility. Emerging technology is also an area that is intangible. For example, it's very hard to predict the next change in technology unless you are Apple, Google, Samsung or Microsoft.

3. Analyse Risks

For each risk you identified in step 2 you need to assess the likelihood of it occurring. Ask yourself if the risk is likely, is it internal, external or random? You are aiming to work out the relationship between the likelihood of it happening and the consequences of the risk in step 4.

We will start by rating each risk against the amount of effort required to manage it, or prevent it from occurring. For example we can use a rating scale of 1 to 4:

1. **Low** – routine procedure required.
2. **Medium** – Allocate specific responsibility to a moderate risk and implement monitoring or response procedures.
3. **High** – This has potentially devastating consequences and requires action.
4. **Extreme** – Requires immediate action.

The result of rating your risks will help you identify those that need immediate attention.

Using a risk analysis matrix makes this task much easier, and allows any trained personnel to make an informed decision based on the risk profile of the company. An example risk matrix is outlined below with information on the likelihood and consequences defined.

LIKELIHOOD	INSIGNIFICANT	MINOR	MODERATE	MAJOR	CATASTROPHIC
A (Almost Certainly)	✔	✔	✗	✗	✗
B (Likely)	✔	✔	✗	✗	✗
C (Possible)	✔	✔	✗	✗	✗
D (Unlikely)	✔	✔	✔	✗	✗
E (Rare)	✔	✔	✔	✗	✗

✗ = Unacceptable
✔ = Acceptable

4. Assess Risks

The purpose of risk assessment is to compare the result of risk analysis (step 3) with the criteria and context of your business (step 1). This results in a decision for each risk on which issues need to be fixed and the priority for the implementation of each.

In order to understand this for each risk, you must look at:
- the importance of the activity this risk affects;
- the degree of control you have over the risk;
- the losses (actual and potential) that could occur;
- the benefits and opportunities that arise from the risk;
- what social factors deem a risk to be unacceptable.

You may discover that you need more information and analysis before you act. It may be that the risk is tolerable, and can be left untreated, or that it is unacceptable to the business or society and therefore something must be done.

5. Treat and Manage Risks

With each identified risk you will want to decide how to minimise it and make changes to either eliminate or limit the potential damage. Each risk will present options on how to treat the situation, and you should look at each in terms of:
- the cost of the treatment.
- whether it is feasible.

- who will be responsible.
- how you can measure the success.

Here are some suggestions on how a treatment of risk may present:
- **Avoidance** – can you avoid the risk by conducting the activity in a different way, but having the same outcome?
- **Control** – can you reduce the consequences or likelihood of the risk occurring (or both)?
- **Transfer** – can you shift responsibility of the risk to another party who is better able to control it or is able to withstand the risk?
- **Retain** – the last option when accepting it cannot be avoided, transferred or controlled.

6. Monitor and Evaluate
Each risk must be monitored and evaluated on a continuous basis. This allows you to understand what is effective in decreasing risk, what doesn't work, and even when management of a risk becomes redundant.

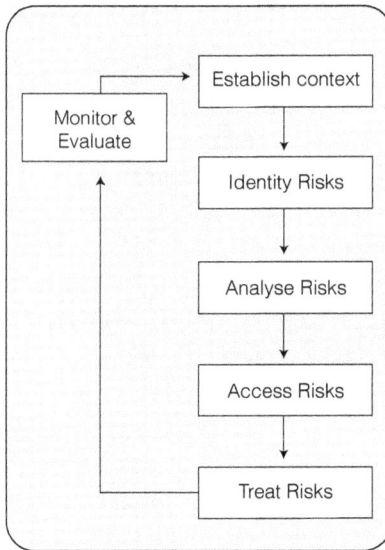

```
                    ┌─────────────────┐
              ┌────►│ Establish context│
┌──────────┐  │     └─────────────────┘
│ Monitor &│  │              │
│ Evaluate │  │              ▼
└──────────┘  │     ┌─────────────────┐
        ▲     │     │ Identity Risks  │
        │     │     └─────────────────┘
        │     │              │
        │     │              ▼
        │     │     ┌─────────────────┐
        │     │     │  Analyse Risks  │
        │     │     └─────────────────┘
        │     │              │
        │     │              ▼
        │     │     ┌─────────────────┐
        │     │     │  Access Risks   │
        │     │     └─────────────────┘
        │     │              │
        │     │              ▼
        │     │     ┌─────────────────┐
        └─────┴─────│  Treat Risks    │
                    └─────────────────┘
```

How is risk management applied in a ski resort?

All too often, resorts apply the latter parts of the risk management process. That is, trying to prevent the risk by laying out markers and alerting people to the immediate problem, and then scooping the injured up once an accident has occurred. There are many risks in a ski resort that simply don't occur anywhere else. Many visitors are coming for their first snow holiday, and only begin to understand the many risks as they improve their skills and spend more time on the slopes.

This partial application of risk management means that resorts miss out on the benefits that good process delivers such as reduction in costs, accidents and better reputation in the market.

Most preparedness exercises are undertaken by individuals or national bodies, which despite the best intentions, both lack the time, knowledge and finance to effectively get the message in front of the people who need it. Ski resorts should get involved with these activities to increase the exposure of these initiatives.

Examples of risk on the snowfield

The snow industry is facing issues around the skill level of customers against the accessibility of terrain and the availability of equipment. Beginners are not educated sufficiently about how a ski run works. Snowboarders do not understand skier movement across the slope and vice versa. They do not anticipate what the other will do and continue to endanger or injure other snow users. Resorts actively promote the availability of higher, riskier jumps and activities to attract people to their facilities. Social media is rife with videos of dangerous and reckless behaviour on the slopes.

Even games and movies highlight risky action as being the ultimate snow experience, they rarely show the background work, and resources required to assess a location for safety. Public awareness of risks and how to mitigate them is not being improved by the kind of hype that encourages the uninitiated to put themselves in danger.

Skiers v snowboarders

The different disciplines out on the snow (skiers, snowboarders, telemarking) are mixed with confined space, resulting in collisions with each other, or natural and man made objects. Each discipline has its own challenges – snowboarders for example have difficulty looking behind them. Each has its own movement pattern, so the shape of the turn and distance required to manoeuvre is different.

When a person from each discipline doesn't understand the other, they will predict the movement based on what they know, not what the person will do in their situation. This is a situation that is more prone to a collision. The issue is prevalent especially on the beginner slopes where awareness to the outside world is lessened because an individual's concentration is focussed on their own learning and fight to stay upright.

Beginners and chairlifts

Chairlifts make it very easy for people to go up hill. When presented by the choice between a t-bar and a chairlift without education a beginner will go for the chairlift. It looks less scary, and there is less chance of looking stupid. However these users don't know what is at the top, and how they might get down. They think about the problem at hand, not the one that will present after it.

So how do we encourage people to become better skilled prior to launching up the hill? How do we ensure these people don't get themselves into a position where they will hurt themselves or someone else because they can't stay in control?

Thredbo sells a ticket designed for beginners that only allows them to use the beginner slopes, and associated chairlift. This is a great way to keep people in the right place, simply because they are not allowed to venture further and are forced to refine their skills. Unfortunately, there is no skills test prior to paying the higher price for the full ticket, so we rely on the customer's perception of their own skill to manage terrain higher up that is unknown to them.

Avalanche

With the passion for adventure growing, people are going into avalanche prone territory more than before. Unfortunately the knowledge required to get in and out safely is not growing at the same rate.

People think that by simply having a beacon with them they will be safe. However these devices are complex and don't all operate well across brands and models. The user also needs to know how to operate the beacon and find someone who is buried quickly. Beacons are useless on their own. To be effective there must be two devices with working batteries, and two well trained people to use them. Time is everything in these rescue situations.

Do you have a beacon familiarisation course in your area? If not, I would strongly encourage that you start one. Get a group of friends together and teach them what you know, share ideas and practice a bit. Setting up a small group of interested parties in an area is really easy and it doesn't need to have a professional involved to start off. You can use Facebook or a service like Eventbrite to publish the event and gather attendees. Once you have a few people coming along, approach your local ski patrol or avalanche centre and ask them to help out with some booklets, guides or even send someone along to have a talk. In the summer do some training by hiding your beacon in open parks, or bury them in the sand at the beach.

Snow pack analysis

What is better than using a beacon? Not having to use one! The key to avalanches is the way layers of snow melt together. Knowing how to identify potential danger from snowfall history, slope pitch, local knowledge and snow pack analysis are key skills that are not being passed on fast enough.

Just looking at the top level of snow is no good as avalanches are caused by layers of snow that are packed on top of each other. The upper layer can be unstable and slide because of its position in relation to the lower levels. When given enough force in the right place, this level will simply slide off the top and create an avalanche.

Skiers must know about the snowfall history and how to identify these risks by digging into the snow to look at the layers. If they are not going to learn this they should get the basic knowledge to identify the right conditions, and stay away from the areas, or only venture there with qualified and experienced help.

Local interest groups, avalanche centres and patrols must share this information, to help people make the right decisions.

Fitness

A large proportion of injuries can be prevented by good physical conditioning.

Skiing and snowboarding for most people are not their regular exercise and so the body is not used to the activity, making it easier to sustain an injury.

Our health plays a large and vital role in how we perform athletically. Best selling author and fitness guru, Mike Campbell, tells us about how the body is affected by lack of sleep, alcohol and poor fitness:[28]

> "If you are not fit, tuned and prepared for skiing or a strenuous activity, your body's physiological response is to protect itself. The majority of your energy will be put towards protection by tensing up and being defensive. Your technique will suffer as a result and you will feel tired faster and ache at the end of the day. Your enjoyment level will be reduced because you are concentrating harder on making your body do something it just doesn't want to do. If you take a fall, because you are stiff your body is likely to sustain a more significant injury because it doesn't relax and bend with the impact.
>
> Alcohol plays a part here too. A vin chaud may feel like it relaxes you, but it slows down your physical ability by thinning your blood and impairing oxygen refill around the body. One of the effects is to make your body feel warm, sending your blood out to your extremities. This isn't the best thing to do in a cold environment or when you sustain injury. Blood would normally be pulled into the body's core to help maintain vital functions. Injury with alcohol can make the injury worse, or at least slow the recovery. This is because your blood is busy in the extremities and your body is also using energy to process

the alcohol rather than on fighting the injury sustained. "

Of course the alpine village environment can motivate people to play hard off piste as well as on. Mike goes on to say: "This adds to negative effects through lack of sleep. The effects are the same as being out on the road the morning after a night out. You are less alert and not in as much control as you would be with sleep and no booze."

This kind of behaviour is of course illegal on the road – you are travelling along in a large metallic object at high speed and the potential danger to yourself and other road users is high. The same risks appear on the slopes and are magnified when many people have done the same thing. Is there a need to discourage alcohol in our villages and mountain restaurants? Possibly, but I don't believe we have the data to back this up, and we must balance the financial implications to the resort experience.

Trained and experienced practitioners can of course deal with factors like these much better than those who are new to the activity. That's because of 'learned memory' where the body's systems know what is required to be able to conduct the activity in a more subconscious fashion. This of course applies to those who have learnt good habits and efficient technique. These people will be able to relax into a ski run relying on their subconscious to deal with 90% of the movement and reactions required. This provides a level of dexterity in performing the fine motor skills required to ski/board well.

Hiking groups and adventure companies specify a level of fitness, or at least a training program, that the customer must adhere to before participating. Should resorts be applying this principle in education to potential customers?

Back/side/nearcountry
Because people are increasingly interested in backcountry adventure, they are forcing out the traditional boundaries of ski resorts. Equipment and lifts are enabling people to take a hike or a walk to these remote areas that would otherwise be too far away.

Alpine touring is becoming more popular with new equipment

making it easier for traditional downhill skiers to walk with skis on. This gear makes them look like telemarkers, but with stiffer boots for use with regular downhill skis and bindings. Snowboarding has a split board that separates in the middle and creates two skis to walk on. This invention has taken away the major barriers of boot packing or snowshoes for snowboard adventurers.

Often these users start in higher risk areas within the resort. These internal areas often have risks mitigated by paid staff patrollers and avalanche control systems such as the 105mm Howitzer used by Telluride Ski Patrol to shell remote areas and force avalanches to occur. If you have never done so, YouTube this, you won't regret it – it puts a smile on my face every time! *http://www.youtube.com/watch?v=WrNekoXniJM*

Markings and signage help the user decide what to do and give them options to return the way they came, or to take an alternate, safer route. But you don't need to go outside the boundary to see an avalanche.

Pushing outside resort boundaries is dangerous. Alaskan areas are finding the increase in numbers skiing the backcountry happening at an alarming rate.

During the past 25 years, Brian Davies, the Snow Safety Director at Eaglecrest Alaska ski area, has been learning about the mountainside in his backyard.

As time has passed, he is seeing lots of folks hiking up out the back, but he is convinced[29] that the avalanche awareness and backcountry skills are not increasing at the same rate. There are many skills required to adventure away from the reach of emergency assistance.

The assumption that passing through resorts' backcountry gate means that it is safe out there is wrong. Appropriate signage must be placed in these locations to alert adventurers to the dangers ahead, how to use an appropriate buddy system, how to get help and advising people to wear a beacon.

Examples of risk management

Ski school
Ski school does a great job at educating students on how to fall properly, all aspects of using a lift, how to get around the mountain and where to find help in the resort. But not everyone takes a lesson!

Courses and knowledge groups
Many resorts around the world are now running avalanche and backcountry awareness courses. Some are more comprehensive than others; Alaska has a great education program. The program's goal is to change the 'know enough to be dangerous' folks into more aware and educated adventurers.

The Southeast Alaska Snow and Avalanche Workshops run in November each year and are built up as a series of 20 minute lectures. This keeps the content accessible, useful and light. The aim is to educate those heading out to the backcountry, and to reverse the trends where deaths and accidents are occurring for people with limited knowledge.

In the well-known snow mecca of the UK, Harry's Avalanche Talks provide a useful refresher and learning curve for those open to learning about avalanche safety. Taking these groups out on to the parks of London they run beacon training through the undergrowth.

Slow pistes
'Area 30' created in 2005 at Grindelwald, Switzerland is an area dedicated solely to resort guests who prefer to leisurely move down the mountain (at no more than 20 to 30 kilometres per hour). It is designed to allow skiers to take their time, enjoy the scenery and importantly, feel safe. This is a new idea in managing safety in resorts, which I personally think is a great idea.

Research completed in Zermatt, which has four dedicated 'slow run areas', shows that accidents caused by collision in these areas has reduced dramatically and policing the speed has not been a problem. A combination of gradient, slow moving skiers and clearly marked speed limit has meant that managing the ski run has not been necessary.

There are now seven of these dedicated ski areas across Switzerland, and the Italians have begun to introduce them to their resorts.

Risk Analysis for a ski patroller

Patrollers are at the forefront of safety for your guests, making them an invaluable resource for assessing and managing risk.

A day in the life of a ski patroller
A typical patroller day consists of a morning briefing, a couple of hours on a sweep of the hill, opening up runs, checking conditions and providing avalanche control.

The bulk of their day is spent assessing conditions and potential risks on the hill. This would include hazard marking, trail work, and assisting other departments with safety. Additionally, they help resort guests with queries, injuries and accidents.

The end of the day consists of a final sweep of the mountain to close all of the runs and to ensure that resort guests are off the hill safely.

Preventatives and Hazard marking

The mountain environment is constantly changing which means patrollers are actively monitoring and marking hazards. To help mark hazards, ski patrol have a number of tools available to them.

Hazard Poles
A hazard pole is a two metre brightly coloured length of bamboo or plastic conduit, used to highlight hazards to resort guests. They are commonly used to mark small hazards like bare patches of grass, holes or bare rocks.

To indicate the area of concern the pole is pushed into the ground or snow between the guest and the hazard in the direction of travel. This forces the guest to navigate around the pole, avoiding the danger.

Poles can be placed individually, or tied together with rope to create a boundary, or rope line fence.

Ski patrollers will also use them to mark drop offs, rough snow and ice, man-made objects (e.g. snow cannons), streams, rivers and more.

Fences

Net fences or pop fences are made out of brightly coloured netting supported by poles at regular intervals. They are strategically placed in positions where there is a danger that a guest needs to be stopped from entering.

This may be a creek line that has a steep entry to it, a physical man-made object like a concrete block, or perhaps a stationary grooming machine.

These hazards are generally well known to all as they are usually large and visible, however guests may not make the connection that there is significant danger.

Rope Lines

Rope lines are made up of a series of poles spread evenly across the area with rope hung across them. The poles can be hazard poles or something more permanent like a fence post.

They are used to mark area boundaries, or to create a chicane that helps direct traffic into a lift entry or busy area.

They are used to mark hazards that do not pose imminent danger to the guest should they cross the line.

Signage

Just like on the roads, signs are used to assist people in making a decision on where they can go and what to expect. Signs are organised into four different categories.

Warnings

Where a hazard exists ahead, or the guest needs to understand the potential risk before setting off, a warning sign will be erected. This will detail the upcoming run, and the risks involved that may cause an accident if a fall was to occur. For example, a steep black run which is generally icy may have a warning sign, which states "If a fall occurs, you are unlikely to be able to stop. You must have the appropriate ability to embark on this run". It would also have a black diamond sign (or a double black according to local protocol) to signify it is one of the hardest runs available.

Other warnings may be used to indicate hidden obstacles. For example at the top of an ungroomed, advanced run where it is not practical to mark every issue below the sign would state "Caution, unmarked obstacles ahead, advanced skiers only".

These warning signs must be positioned in a location where every guest who enters the run passes by it.

Trail Markers

These signs are used to assist in navigation and help the user decide which route to take. Most resorts use the international standard grading system of green, blue, red, black and double black diamond to show the level of skill required. The sign may also represent a shape that is coordinated with the colour, so a blue may be square, red a circle and black is diamond shaped.

Navigational Markers

These are heavy duty hazard poles, which are placed semi-permanently to help guests navigate along a run in poor visibility. They have reflective material on them to make them visible at night and in bad conditions. The distance between the poles is set so that when you stand in the centre between two markers you can see both. This way a guest can move from one to the next in whiteout conditions and have a way to navigate the run safely.

Informational

Informational signs cover temporary messages such as lift closures through to more permanent messages about lift and trail status. They are placed in high traffic areas to ensure resort guests receive the information. In some resorts this information is being communicated via local radio and smartphone applications.

Risk Management for a patroller

When a potential risk is found by a patroller, either in the course of their inspections, or because of an accident, they have to decide which action to take next. How the patroller arrives at that conclusion is affected by a number of factors:

Experience
The patroller's accumulated knowledge and individual experiences play a big part in the way that they make decisions. The number of years in the job, the previous accidents they have seen, and education they have received helps them to make better judgement calls.

Accident Information
Understanding previous accidents and resulting injuries in that area will affect the level of action a patroller takes.

Risk Matrix
As previously covered in this chapter the risk management matrix provides an excellent tool for a patroller to set a preventative action.

This matrix helps patrollers decide what to do within the resort's defined health and safety guidelines without needing to know all the legal complexities.

Many risk management decisions are low level and are dealt with quickly by the patroller. For example a bare patch of grass in the middle of a run that is obscured by the terrain can be marked with a single hazard pole above the area, and so can be clearly seen and avoided. Other situations are much more complex in their nature, for example tree wells.

A tree well is a void or depression that forms around the base of a tree, containing a mix of low hanging branches, loose snow and air creating a confined space. These voids can be hidden from view by the tree's low hanging branches making it difficult for skiers to be found if they fall in, or to self rescue.

On these runs people are encouraged to ski with a buddy. If an incident like the tree well unfolds, they are able to assist each other immediately, reducing the chance of serious injury.

Although the occurrence of serious injury is rare, tree wells are common across many alpine environments and present a complex problem for patrollers, especially when it comes to marking the hazard.

A patroller will consider the following questions before making their decision:

1. Mark only one tree well – the largest one?
2. Mark all tree wells on the run?
3. Mark no trees and assume the skiers know the risks?
4. Put up an informational sign at the entry to the run?
5. Close the run?

Each of these options present practical, social, liability and legal challenges. A patroller will make his or her decision based on experience, training, and resort risk management protocol.

These decisions may come easily to the experienced patroller, but you must ensure that every person has the ability to make the right choice. Training and gaining experience on the many different scenarios can take years.

On a regular basis as a normal part of their job, ski patrollers assume levels of risk that would be considered unacceptable in many other workplaces.

The death of Bill Foster, an Alpine Meadows, Tahoe patroller, highlights this graphically, in a tragic tale. Bill was undertaking routine avalanche control work in December 2012 when he was standing in the "safe-spot" that had been used hundreds of times before. After 30 years as a patroller at Alpine Meadows the avalanche hit him that day, and took his life.

Unfortunately the standard procedure that had been developed and followed to the letter, day after day, failed. [30]

Terrain control tasks such as ski cutting of wind lips, and cornices, in big terrain (just one example) is literally an unacceptable level of risk by any normal standard of workplace safety and there is no adequate mitigation of the risk to change that. Operating in a near zero visibility whiteout, where you can barely see your outstretched gloved hand is not uncommon. In other workplaces the unacceptable risk would mean the work is stopped immediately, but you simply cannot run a ski area and open the big stuff without this work being done.

In a day's work a patroller will perform multiple and diverse roles. They work as a search and rescue expert and dog handler, a medic, explosives handler, snow scientist for avalanche control, not to mention the grunt work involved in setting up fences, signage and rope lines across the resort. After all that they answer customers' questions and represent their resort in a dignified manner.

Most patrollers do this work for pitifully low wages because it's just part of the job. They are considered unskilled workers, even though many of them hold Emergency Medical Tech (EMT) qualifications and licenses to handle explosives. These are qualifications that take many months of study, practice and exams to pass.

Patrols around the world adopt safe practices wherever possible, such as using rope techniques when cutting cornices, safe spots below rocks for avalanche control duties and explosives control. But this doesn't change the fact that the men and women who are up on the hill before dawn and down after nightfall do it all so selflessly, tirelessly, for the love of the sport and for so little in return.

This work is critical to the enjoyment of so many customers, and ultimately comes back to resort reputation. It saves lives.

This is a situation that is understandable from a commercial point of view. In the USA there is no job description for a ski patroller, and the job is classed as unskilled labour, requiring only minimum wage, and pay rises as defined by the resort. For those looking to commit themselves to the year round winter, this means that gaining a visa is almost impossible depending on the country you are coming from.

Is it fair? No, not really. But I believe ski patrollers are a different breed to any other because of the bond, selflessness and willingness to help others. Ultimately it would be great for the work of these men and women to be recognised by the governments whose jurisdictions they work within, just like any other profession, thus providing avenues to better pay. Generally though, it's not about pay for anyone in a ski resort, it's about the lifestyle, the camaraderie and the outdoors.

Ski patrollers are driven by a moral purpose, rather than financial. So please, honour these valuable people on your staff with the uniform and the equipment they need to be safe while operating in the harsh environment they do. It's not right that they should have to pay for a vital piece of equipment that will help them save a patient's life on someone's payroll. Your patrollers should also be recognised, and a central part of your resort image. They should be respected, not feared, maintaining modesty and integrity whilst representing your business.

Communicate & Reassess

Senior patrollers and risk managers carry out the review process for accidents. They will use the collected data, training and personal experience to assess the impact of accidents to the resort risk profile.

The reassessment process will analyse the issue using your defined risk assessment matrix to decide how likely an incident could reoccur and weigh that against the information about what is acceptable risk. This assessment is critical to undertake because a future accident in the same spot would raise further issues of liability, especially if there is no documented analysis of the event.

The results of this assessment may bring up new ideas about the preventative taken or changes required to the terrain, and may result in new process or procedure. This data should also be used as a critical measure to see if you need to tweak your risk analysis matrix because social, legal or liability factors have changed.

In social terms, the acceptable level of risk changes because of liability and guest experiences, the media and the views of society. When assessing risk, remember that we are in the business of experiences. It's not good for business when guests are thinking they are in a 'nanny state' – the very nature of a ski area provides a level of hope for action and adventure, something to get the blood pumping for our guests.

Outcomes such as any change to policy or work procedure as a result of an investigation should be taken back to your staff through team meetings, to patrol and through to mountain management.

Everyone needs to know what is being done to help your guests have a better experience. Education by example and understanding of the environment encourages staff to identify risks, before they become serious.

Workbook / Tips

- Write down the process you use to assess a risk on the hill.
- What area are you most focussed on when making that assessment (guest, resort reputation, financial, insurance)?
- What education could you put together to help beginners be safer?

Data

Your information is collected everywhere: when you visit websites, when you're at the supermarket, travelling, in hospitals and hotels. This data is used for analysis, in most cases to try and understand you and find new ways to help you spend your money. Some is collected for academic research to draw conclusions about why things occur, and how humans think.

Data collection has a similar importance for the leisure industry. In a ski resort, data tells the operators where you go, how many lifts you take, what equipment you rented and what bar you drank in. They *should* also know what injuries you sustained, why it happened, who helped you and what was done to prevent it happening again.

If a skier attends the medical centre, they complete a form. What happens to that data? Is it used in the best manner? How much of it actually goes back for analysis to work out how to improve the resort? Why are we, as an industry still stuck in the paper-based world and not using mobile devices?

Perhaps the worst thing in the industry right now is the lack of attention given to people's mishaps by resort management. The litigious society we live in results in paperwork as an act of protection, not of being proactive in managing safety.

By collecting data and using it in the right way, we can transform our industry into one where we can be in control - an industry where we can act with certainty and measure the results, and where we can truthfully say we have done our best.

Data recording is essential for good risk management process and for the prevention of accidents. If you have data on accidents but cannot compare that to preventative measures you have made, then you cannot tell if the work you are doing is actually making a difference.

For example, in avalanche prone territory, data is one of the key tools to keeping people safe. Information about snowfall, snowpack and weather are built up over time to help predict slides and take appropriate action prior to an incident.

In fact, there is now so much data about avalanche prone terrain that there are dedicated avalanche centres all over the world manned by specially trained staff.

Data collected by the avalanche centres is not only used to protect the patrollers and public who enter this terrain on a daily basis, but it is used to identify longer term trends and back up short term assessments. Larger volumes of data mean it can be harder to analyse, but with the right tools it can be used very effectively to predict dangers.

Collecting accident information and matching trends against preventative actions help patrollers make good decisions and reveal if their decision-making is effective.

This combination of data helps if an incident were to escalate. You will know that a trained staff member has visited that scene, made use of the data available and made a call about any work that should be carried out to prevent further occurrences. This information is vital to prove in a court case that you have done what is required and made the best effort to improve the area, and that the situation can be learnt from.

As an industry there is an inconsistency of participation and accident data, resulting in critical questions being raised, such as those posed by Dr Tracey J Dickson in her book "Risk Management in the Outdoors". When drawing on publically available data in the broader outdoor pursuits sectors, that includes snow sports, she notes:

"There is
- *inconsistent or no definition of what may be a reportable injury across various sports and activities*
- *inconsistent definitions of participation or exposure*
- *a mix of voluntary and mandatory reporting which may lead to under or overrepresentation of injuries*
- *a lack of data on injuries that did not result in hospitalisation which may provide indicators of potential problems."*

Dr Dickson's findings were gathered across data from United States, Canada, Australia and New Zealand, and my interpretation of her statement is that, not surprisingly, none of them "have their act together".

This is not just a high level national issue. Resorts struggle to get a good picture of the data and reasoning behind events. This unwillingness to share findings results in different collection methodologies and isolated learning in resorts. You have the power through this book to make a change that will help researchers like Dr Dickson provide more insightful conclusions by removing some of the data issues we face now.

Effective medical treatment relies on making sure the right people know the right things at the right time. Resorts currently rely heavily on dispatchers to collect information to help coordinate their patrollers. To help deal with their workload many resorts now use computers to collate information – which is great.

However, the quality of the data collected at the moment is only good enough to cover statistics like workload, tracking of events and high level demographics. It does not help in the transfer of a patient to the next level of care, or in proactive risk management, because there is not a great level of detail that is accumulated from patient, to patroller to dispatcher to handover.

By using a smartphone for reporting accidents, we can collect a location for each. This helps to cut down response time, and provide useful analytical data for reporting.

There has never been a better time to use technology to increase your knowledge, and make your customers safer.

As an operator in the industry, you have the ability to save lives.

Data Collection

Data is crucial to improving your business. One of the main goals for a ski resort is to make it safer. Offering a safe environment improves the guest experience.

Having access to more data allows us to be better prepared; to understand the trends of our guests; and make better decisions on our trail design and hazard points.

Not having data results in inaccurate assumptions being made about skier safety, potentially resulting in incorrect resource allocation and ineffective preventative actions.

> *"Assumptions are the mother of all mistakes"*
> – Eugene Lewis Fordsworthe[31]

The current accident reporting process in place in many resorts involves a paper and pen effort filled in by a ski patroller. Another patroller will then review the paper report and determine what other actions must then be taken. In many instances this information is then double handled and transcribed into a computer system.

This process is cumbersome and open to mistakes.

Current issues with data collection

As mentioned ski patrollers currently handwrite their reports, and there are many factors that make this process fallible.

The Human Element
The human element impacts on data collection. In the heat of the moment, a patroller will be working at a high rate to get on top of patient care. Factors that result in poor reporting include tiredness,

intensity of the accident, experience of the patroller, the kind of incident attended, time available to complete the report well, and perception of the situation.

Additionally, bystanders and witnesses will suffer from the same issues when they witness traumatic circumstances. This is particularly the case if they are related to the accident victim. Feelings of despair and panic set in and distort their perception of the event. This also leads to people not noticing if information is missing and a full recollection could take weeks or months to surface if the event is traumatic enough.

This can lead to wildly varied and conflicting reports from each of the parties involved.

Garbage in, Garbage out

It is important for the patroller to be invested in the process of collecting data. They need to know the full extent of what happens to the data, and why it is being recorded. If they know that the data they record is fully utilised to learn and improve their world they are more likely to provide better information.

In written forms, this translates into the patroller writing neatly on the form, and spending the time to fill all the fields out with detailed information. They must use their training to remember what fields are required, and what information goes where. Sometimes fields must be linked using a numbering system, and so training on filling the form is required.

On a digital form, this translates into less training and less mistakes. The system will guide the user through the form, tell them what is required, and give them contextual feedback as they move through it.

Geographic Accuracy

Another major problem with handwritten reports is the inability to accurately describe a location. Typically a location might be reported like this:

"*Top of Excalibur, on the right by the trees.*"

Just to fill in the blanks, Excalibur is a fictional, but typical intermediate, wide, on piste run. It is densely wooded on either side from top to bottom, with a tree island in the middle.

As you can see this simple statement regarding location is not detailed enough.

Assuming we get this correct, how far down the slope do we need to look? We could possibly begin in the top 50 metres, but that is still a big area. How close to the edge is it located? Two or 20 metres?

Can you imagine what a lawyer might want to know when investigating the accident? How much doubt could counsel introduce with a description like this in court?

Is it time for a move to smartphones to capture data quickly and accurately?

Hybrid data systems don't work!

Double handling of data, from a paper system over to an electronic system increases the possibility of data entry mistakes.

The factors that introduce these mistakes are
- misinterpretation of handwriting
- mis-keying on entry
- investment of the person doing the entry

If a patroller were to use a smartphone to collect the data there is a greater chance that the information is going to be accurate the first time around, as it eliminates many human factors.

However, in resorts that employ a hybrid approach from paper to computer some data is lost in the transfer process. They limit the information that is transferred to save time and only move the data that is essential to reporting.

This leaves many useful data points on paper that would never be able to be used in an effective manner, and once that page is turned the data is dead. Direct entry through a smartphone ensures that all data collected becomes available for analysis.

In the event of litigation having all the data in one place cuts down the time to find and compile any evidence required.

Accident Investigations

Accident investigations are undertaken by a team of specially trained patrollers who understand the requirements set down by law. These investigations have to be conducted because there is insufficient data recorded at the time of the event.

Incidents that occur with man-made objects, major injury, rental equipment, ski school class, vehicle or involving other members of staff, are usually subject to further investigations. This is because relevant authorities need reports to approve insurance, verify staff injury, or replacement of equipment. The courts require evidence to make a judgment about who is liable and where repayment will come from. These cases can be complicated and take time to resolve as there are multiple parties involved.

The guidelines in some USA resorts are so all-encompassing that about 60% of incidents result in an accident investigation of some sort.

To complete an accident investigation, specific equipment is required – a good quality camera, forms, paper to draw diagrams, and even a Global Positioning System (GPS) device to log the exact coordinates. Investigations usually involve locating the exact point where the patient was found, and mapping out their path and mechanism of injury. The GPS provides the exact location for future reference, and photos record key landmarks. Supporting diagrams are drawn to illustrate the timeline of events.

All of the information is then compiled with the patroller statement, witness and accident reports.

In order for the photographs to be accurate and represent the incident well, these investigations need to happen as close to the time of the event as possible, preferably with the patient in the pic-ture. This would require the team to be available when the accident occurs and be equipped to do so.

In reality most accident investigations happen after the event so the area ends up looking a bit like a crime scene with the chalk man on the ground.

The investigations all take different amounts of time to complete and require people with differing levels of skill depending on the mechanism of injury and complexity of the incident. However this takes a significant amount of work, resulting in trained patrollers getting caught up in accident investigation and doing very little real patrolling.

Big Bear Resort in California has developed one of the most advanced accident investigation procedures in use today. They have devised a system that allow patrollers to identify what level of investigation is required through five radio codes:

1. **10-64:** a few pictures and simple forms are filled out on the scene
2. **10-65:** as 10-64 plus diagrams and patroller, patient and witness statements
3. **10-66:** as 10-65 plus risk management team for independent investigation
4. **10-70:** lift accidents
5. **10-71:** lift accidents plus risk management team for independent investigation

Although this is a simple procedure, the problem is that it gives the patroller even more to think about during the accident. On top of patient care, they must also assess the incident scene and decide which level of response is required based on the possible legal outcomes and patient recovery. They must remember the code because broadcasting the exact nature of the injury over a radio is undesirable. Radios are near customers all the time, and the conversation could be overheard.

Due to the nature of the accident, time taken and resources available, these accident investigations may not occur on the same day. This can often result in distorted or lost information.

Many of these issues can be addressed with smartphone technology.

Using a smartphone on scene allows a patroller to collect infor-mation quickly and easily. It guides the patroller to gather the correct information according to the nature of the accident, reducing the need for and amount of work required in accident investigations.

In the event that an investigation is still required, the data from the smartphone is a valuable tool in supporting the subsequent investigation.

Scene Management

Let's look at some of the ways the data collection procedure on a smartphone can cut down time and increase the viability and quality of the data collected.

Arrival on scene

When the patroller uses the phone to start recording a new case, it will log the time and GPS coordinates of the accident. The patroller may be unable to collect further information because of the patient status or weather factors. Therefore the device can be placed back in a pocket for the duration of the treatment if no opportunity arises to do further work. The remainder of the data collection can be done later when conditions allow.

Witness Statements

While treatment occurs, the smartphone can be handed to bystanders who can create a simple voice or video recording. This timely capture provides an accurate version of the event. There is no delay of time that clouds the witness vision, or outside factor to influence their view.

The witness can be presented with instructions on what to say, which can be defined by the legal jurisdiction under which you operate. You may need witnesses to state their name, address and date of birth in order to be able to identify them. This can be followed by words such as: 'this is a true and accurate account of the event I have just witnessed' before they carry on and tell the story.

Technology can now take the sound track of the report and convert it to text, saving the patroller and the witness time. The audio version can be replayed at any time, even in court if required. The text version of the event will provide great benefit in data analysis later by utilising keyword research.

Photographs

Many resorts equip patrollers with disposable cameras. These are used to take photos on accident scenes as primary evidence. They are stored and developed if required for legal or insurance purposes.

As useful as disposable cameras are, they can present a few issues for patrollers.

In addition to the challenges of deteriorating film or poor quality images, patrollers often struggle to remember exposure numbers when writing out their accident reports.

Photos from accident investigations can be misleading or incorrect because they are often taken after the event when the environment has changed.

Smartphone cameras offer very high quality images that are time and GPS location tagged. The patroller has the benefit of reviewing the image for quality prior to saving on the accident report and can take as many as required. Any photos taken at the scene will show it in real time, capturing the environment (weather, snow condition, patient position) accurately.

The patroller can also take high quality videos of the scene, creating a detailed and useful recording for later use.

Electronic Signatures

Witnesses and patients can sign the screen on the device with their finger and produce an electronic signature that can be printed later if necessary. As technology has advanced, this form of a signature has become acceptable identification/authorisation by the law courts and/or government.

Medical Data Accessories

There are many pieces of equipment available to supplement the functionality of a smartphone that build a picture of the patient's welfare at the accident scene.

One example is a fingertip pulse oximeter. This is a small portable device that measures and records patient's pulse and oxygen levels.

Adding a pulse oximeter that only weighs a few grams and fits in the palm of your hand to a patroller's equipment will help improve their ability to assess a patient's medical situation, and assist with decisions regarding the next level of care. Furthermore any data collected will be useful for the doctor, medical centre or hospital in their assessment and care of the patient at handover.

Connecting the pulse oximeter to a smartphone enables recording and communication of this critical information at the scene, reinforcing the chain of care significantly.

Patient Care

Using a smartphone on scene provides faster and more accurate communication and monitoring of the patient status, as well as comprehensive information for later analysis.

There will of course be instances where the use of a smartphone is not viable when treating a patient – such as where a patient is unconscious, bleeding or requires CPR and there are no extra patrollers available.

Here the primary focus for a patroller is patient care, and they are not expected to carry out other tasks.

Clearly this is not an everyday or every-accident scenario. However by making sure you collect as much as you can on scene in the most efficient way possible, we will have better coverage of all accidents and not just the ones that look like they could become a legal issue.

Communication and data is key to efficiently managing the resources of the patrol. In order to complete jobs (e.g. fencing) and respond quickly to incidents, someone must coordinate the patrollers and equipment with knowledge on their location and availability.

The coordinator of the team needs regular and accurate updates on the situations patrollers are dealing with and their locations. This information is critical to smooth running of the resort and efficient response to any event that comes up.

This is manageable on a hill with a team of 20 to 30, and one coordinator. Beyond that, it is difficult to keep up with equipment, locations and jobs on the go without a system to help record the information.

Dispatch systems provide a central communication point for this exact purpose. All patrollers communicate with dispatch and receive instructions from there.

The dispatcher records all conversations and actions in a time log, and liaises with other services to organise handovers or extra assistance.

The dispatch system can also be used when data is received from the scene. In the preventative phase, decisions can be made on the provision of an extra resource to the scene to assess the impact of the incident. This call needs to be made by the senior patroller on staff, or the dispatcher who can see more information about the incident from afar. They will also know who is closest or available to attend and make that assessment. This is an important step in ensuring that the cause of the accident can be addressed and made safer as quickly as possible. As we will discuss soon, having the data is not enough, you must act on it when an issue is spotted.

Communication of Patient Data

Once contact is made with a patient and protocol is followed, there are two outcomes. The patient will either decline and walk away, or the patroller will treat the patient and take them to medical care. Both require paperwork as evidence of the process.

The difference is that a decline of care is processed on the snow, on paper, in challenging conditions. The information the patroller collects during a treatment is collected in their head and on paper. The detail of either scenario is not fully apparent to anyone but the patroller until a full report is completed after the patient has been released to the next level of care. Both situations take the patroller out of circulation to complete paperwork, and neither provide good communication to supporting staff.

While the patroller starts to assess the patient and decide what treatment is required, time is ticking away. This assessment usually results in one of two outcomes:

1. Decline of care

In the case where the patroller makes an assessment and the patient is well enough to proceed on their own, they have the option to decline care. This however presents an issue in regard to the patroller's duty of care and breaks the chain or protocol of transporting the patient to the next level of medical help.

This also puts the patroller in a position under which, should the patient go on and have a secondary accident, they could accuse the resort or patroller with negligence. This must be mitigated through the patient's acknowledgement that they are satisfied, understand that they continue at their own risk, and are choosing to turn down medical aid.

Using a smartphone to complete this task removes the double handling of data between patroller, paper and a report that is created later. The device can collect the information on scene, and allows the patroller to stay available for another job rather than being chained to a desk filling out a report.

2. Patient Treatment & Transport

If the patient is incapacitated or decides to take medical treatment, the patroller will order supplies and equipment according to their injuries. This may include: more patrollers, specialist spinal equipment, pain relief and a sled.

While the patroller is waiting for equipment to arrive, they prepare the patient with anything they have on their person as well as making space for the transport. They will also spend this time conducting secondary assessments and in-depth checks to ensure nothing is missed.

With a serious incident, radios limit the patroller from stating the exact nature of the injury. The reasons for this have already been discussed: it's not good practice because a member of the public could overhear the report.

I have been involved in incidents where the full nature of the injury could not been communicated in detail over the radio. An experienced listener can however hear the tone of voice and pick up that there is stress and tension in the information, and respond accordingly.

This however is not enough. In resorts where a doctor is on the hill, or close by in a medical centre, they can only make the decision to attend if it is well communicated. A doctor on the scene of serious trauma can make a significant difference to the outcome of the patient. They are able to provide drugs and assess injury to a much more advanced level than that of a patroller.

If the doctor is not available and the injury is life threatening, recording data as it happens is critical for the chain of care and debriefing process of an incident.

Smartphones provide the technology and connectivity to record and communicate the patient's situation to the next level of care. This will provide information ahead of time, allowing better preparation prior to the transfer of the patient. This live transfer of detailed data also allows remote viewers to provide input and support if they are unable to attend.

It also allows the patroller to complete their paper work once, with no double handling, and get back on hill quickly.

A patroller that uses a smartphone on scene in this manner can use any time waiting for assistance or equipment to start collection of data.

However there are challenges with balancing patient care with data recording and communications. In these situations a dedicated data recorder provides an invaluable resource on which to rely, while the responders and professionals dealing with the event can concentrate on providing the best care, and executing the best plan to help the patient.

A mobile phone can collect and keep this data accurately and safely from the start of the process through to handover to doctors or ambulance personnel involved in the chain of care. We will discuss this idea, and its implementation in more detail in the future chapters. However, I need to point out one thing that is key when on scene with a patient:

Communication

Smartphone apps allow us to communicate incident data in real time to multiple subscribers such as the dispatch operator, the doctor, mountain management or the local hospital. This provides an amazing opportunity for the next person in the chain of care to be prepared and have all the information they need to get the best possible outcome for the patient.

Databases

Data collection methods allow us to gather a body of data that can be analysed to find trends and patterns. However, even once the conclusion has been reached, the researcher must look at the way the data is handled through its lifetime and determine how accurate it is.

Evidence is powerful

Data collected from one event can be used in many different ways that are secondary to the source of the data.

CASE STUDY 1: RED BULL

Take a look at Felix Baumgartner's recent descent from 24 miles above the earth.

He hit speeds of 833.9 miles per hour, over a 4 minute 20 second descent. The entire adventure was to do more than create media hype for Red Bull and break a few records. The project had other goals – to collect data for research into future developments in space suits, high altitude parachutes, and expand our knowledge on the effects of high altitude and supersonic speeds on the human body.

The data was collected by *National Aeronautic Association* and *United States Parachute Association*, and will be analysed and verified by international organisations, the *Austrian Aeroclub*, and finally validated by the *Fédération Aéronautique Internationale* for ratification of the records he has broken.

One of the biggest benefits of this project is to understand these effects on the body when a high altitude bail-out occurs. This is useful in the budding commercial space flight industry. The risk they are undertaking is huge, and largely unknown. They must find ways to prove to their potential customers, financial backers and insurers that they have the systems in place should something go wrong, and to ensure every possibility of a positive outcome.

Data collection from this record breaking endeavour will ultimately contribute to improved space suits, and advances in technology and medical application in this field.

CASE STUDY 2: SURF LIFE SAVING

Look at the work of Surf Life Saving Australia (SLSA) and check out their incredible decrease in drowning deaths over the past four years covering more than 25,000 kilometres of coastline.

Consider the amount of coastline they are working to protect and the millions of people on beaches in Australia each year.

They have changed the way risk is managed and pulled the data into public awareness campaigns. This has been leveraged to build strong relationships with tourism and changed the way people behave on beaches, ultimately preventing death and injuries.

Mobile Risk Analysis

In 2011, the organisation embarked on a project to change

the way it worked in analysing the risks on beaches and the prevailing conditions. With the convergence of technology into one mobile device, SLSA saw the opportunity to change the way this is done and move away from multiple pieces of equipment and complex paper-based forms.

The end result is an application that collects lots of data about the area, environment and conditions and helps the user make decisions on the situation being assessed. It has resulted in a threefold efficiency gain, and revolutionises the way in which SLSA and land managers assess and manage the risk along the coastlines for which they are responsible.

Accident database

SLSA collects incident data from its own Surf Guard Incident Report Database (IRD), the National Coronial Information Service (NCIS) and by monitoring media reports for drowning incidents. These incidents are recorded manually and on paper with a wealth of information collected and then transferred into a database.

The following information is recorded for each drowning incident:

State; date; drowning location; GPS coordinates; time; age; gender; incident type; activity information; whether the incident was work related; entered into IRD; IRD number; NCIS case number; whether the case is open/closed; whether the case was reported by the state; the original source of information; drowning location suburb; local government area; postcode; associated SLS club; month; day; season; quarter; victim's name; address; residence country; residence distance to coastline; residence distance to drowning location; victim's birth country; nationality; time in Australia; main language; additional activity information; the victim's experience in the activity; whether the incident was rip current related; detailed description of the incident; details relating to alcohol, drugs, or health conditions; weather conditions; wind conditions; sea conditions/wave size; wave type; water surface; temperature; tides; location to a lifeguarding service; whether the location was patrolled at the time; personnel who first sighted the incident; first rescued; other services involved; and resuscitation details.

This data is then analysed by the national risk management team at SLSA Headquarters to help find and develop ongoing strategies. These findings are used to shape policy and procedure and change the direction of public awareness initiatives.

Human benefit

Through the data collected and the initiatives put in place to deal with public awareness, communications, patrol education, data collection and analysis, the 2011 report cites the rate of drowning deaths in rip currents is 50% below the seven year average of 0.10 deaths per 100,000 head of population.

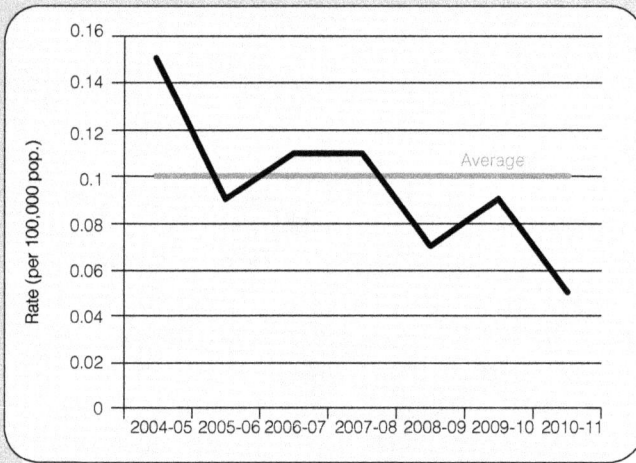

National Coastal Safety Report 2011, Surf Life Saving Australia

Economic Benefit

Price Waterhouse Coopers' (PWC) economic report on Surf Life Saving puts their value at $3.6 billion[32] in 2009/10. This is based on their output-based approach to measuring value in terms of lives saved and injuries avoided as a result of surf lifesaver and lifeguard activities. The flow-on effects to the broader Australian economy equates to an additional $154 million, where injury-prevention allows increased consumption, increased industry activity, and additional labour supply.

That's a lot of money for a volunteer organisation that puts in $42.4 million worth of man hours.

The cost-to-benefit ratio of Surf Life Saving Australia is 29.3 to 1.

Why is this important?

We are in a place where more can be done to benefit the public and our customers; to benefit our companies and organisations; and to the industry we all work in and rely on for our livelihoods.

If we can provide better data then we can benefit as a community and industry.

Researchers like Dr Dickson, who have access to extensive data, will have more to work with and this will provide the ability to draw more informed conclusions.

Businesses, such as the manufacturers of ski and snowboard equipment, helmets, and safety gear, will be able to understand trends on a much larger scale and develop better gear, potentially at a lower price.

Insurers like Robin Barham are able to see that we are doing everything in our power to work on the safety of our guests, and reduce the risks in front of us on a daily basis. This will allow him to provide a better risk profile and lower your insurance costs.

With the accumulation of more data comes greater responsibility. Instead of analysing incidents when the data has been put into a computer from a book sporadically during the season, electronic data capture makes valuable information available immediately.

You may be reading this book because you recognise the opportunity technology is providing us, or you recognise that your patrol can do more to assist in the profits of the organisation and in your guest experience. I suspect that means you want to change your systems and make improvements.

If you collect detailed accident data and the information is available to you instantly, you have a responsibility to monitor it. If you spot two or more accidents in the same place, you should be taking steps to analyse that area and make changes to prevent further accidents.

Insurance companies and the legal system will want to look at incident trends at your resort location. If you know there have been a string of issues in one spot, and you did nothing about it; you are potentially liable for that issue.

Make your data collection as good as it can be by leveraging technology to help collect it. The data can help you directly, but can also be useful for the industry as a whole. The better the data, the better our conclusions will be through analysis and research.

CASE STUDY 3: FEDERATION INTERNATIONALE DU SKI – INJURY SURVEILLANCE SYSTEM

The Federation Internationale du Ski (FIS) Injury Surveillance System (ISS) started collecting data in 2006. This system is designed to provide a reliable way for all FIS teams to report their athletes' injuries during racing and official training. This allows the FIS to collect a body of data from which to conduct research and draw conclusions.

"The next step is to find out how we can reduce these injuries. To do that we still need to improve our understanding of the causes of injury, their risk factors and injury mechanisms using techniques such as video and biomechanical analysis. This information is critical to be able to develop effective methods to prevent injuries."[33]

The FIS is committed to understanding how the athlete's injury occurred and to find ways of helping prevent that injury from reoccurring to another elite competitor.

This system relies on the team doctor, or race doctor, to fill out a second accident report form and return it to the FIS ISS team in Oslo. It's not bullet proof, but the outline of the system is very clear and the engagement to assist elite athletes in achieving their best performance is apparent on every level.

This data collection, right at the coalface of the professional race circuit, provides essential information upon which the organisation can act.

This data can also change the way manufacturers design race equipment. If information is shared in this way it provides a holistic approach to the safety of the participants.

Take the change to the guidelines for the minimum radius of Giant Slalom skis in 2011. They are now longer than in previous years (not as long as straight skis of old), making the minimum radius 40 metres, up from the previous of 27 metres.

This is a change based on research the FIS undertook to provide a greater level of safety by slowing the skier down with slightly larger skis. The turn radius is longer, making it harder to get around the gates in a Giant Slalom race, and thus the athletes must skid a turn more often, washing off speed.

Thought leaders and academics mostly rely on surveys and questionnaires to gather data upon which to carry out research. Large scale surveys are hard to carry out as they can require a significant investment and participation. These surveys are easier to carry out with an authority like the FIS as the potential outcomes can help to encourage participation.

Why should the elite have all the benefits? Let's make a model like this and apply it to our customers, and help them have the best time of their lives! Better yet, let's collect the information from the source, rather than after the fact, in a situation where the data may not otherwise be forthcoming.

Qualitative & Quantitative data

Once data has been collected it is important to make use of it.

You already have quantitative data such as participant numbers and accident occurrences, which will allow you to be able to see things like accidents per skier days. Just having one of these does not allow you to make comparisons or judge the trends over time.

You will also need qualitative data on the issues that may cause an event because this data is interpretive and seeks the meaning of data via words or pictures. This could be the dynamics of the group, or effects of the surrounding environment on the participants. Other factors for an event may be dehydration, lack of sleep or food, and drugs or alcohol. This type of data allows greater reflection against raw numbers.

Let's look at a few of the common data collection techniques in use today.

Surveys

Ask people to voluntarily submit answers to a set of questions. The survey must be carefully designed with the required data in mind. The wording on questions must be explicit and ensure the subject responds in the appropriate, accurate manner.

Surveys don't always work, especially if there is no carrot available. Members of the public need an incentive to complete the work, when they do not see a direct benefit to them. Even professionals will be hesitant to provide their time to give thoughts, as it may impact on the more immediate issues in their To Do lists.

Let's look at how researcher Dr Tracey Dickson had to change methods to be able to complete her work.

"Questionnaires had been placed with the ski patrol, resort medical centres, as well as the doctors, physiotherapists, masseurs, and telemark equipment suppliers in the local feeder town of Jindabyne. After more than a month, there were less than six injury reports even with regular follow up. The number of reports was lower than the researcher expected so a shift was made to begin to understand who was

telemarking, to try and explain why injury rates appeared to below. Participation questionnaires were distributed through telemark events conducted in Perisher Blue, at a telemark film event in Jindabyne and through a retail store specialising in telemark equipment." [34]

This shows how the design of the questionnaire, its delivery and method of targeting can change the response rate dramatically. Any increase in workload for the participants would be shunned unless the perceived value to them was very clear. However there is a lesson here too, that the rate of injury during the time of collection for telemark skiers was incredibly low. The change in the collection approach was initiated to change the understanding of why it was so low, thus targeting the niche and places of interest to that niche. This enabled the content and questions to be very focused and the results were very different – 88 surveys were completed.

Interviews

Sitting with a person and asking them questions is time consuming. Once you have decided on your questions, you must make the appointment, do the interview, which may take other paths depending on the answers, and then transcribe the whole thing, and reread for other nuggets of information. This approach is not suited to raw data analysis as it does not provide you with data in a specified format, rather it provides opinions of people that must be qualified and crosschecked.

There are a number of sayings about getting feedback on ideas. One goes along the lines of *'You will never get negative feedback from friends and family'* and another *'Every idea should be validated by 10 strangers'*.

So in order to ensure that any data coming out of an interview is correct, and can be relied upon, answers should be crosschecked with a number of other interviewees and information databases, which can provide data insights to the question.

Collecting qualitative data is just as important as quantitative data as it provides some perspectives and angles that would not otherwise exist. This is why we take photos on an accident scene as visual evidence. The photo will give the viewer a better understanding of the environment, conditions and help them relate better to the numbers.

Securing electronic data

Information is power. As such the way you handle the information you collect is crucial, and keeping it safe is your responsibility.

In 2012 we witnessed two high profile security breaches at Twitter and Facebook where a substantial amount of personal data was stolen. In 2013 Adobe lost 150 millions users' account information in a hacking event. In instances where personal details such as email addresses, passwords and banking details are stolen, they can be used to steal identities and rob people of money.

Using smartphones to collect and communicate data requires a secure cloud-based system to join all of the devices and make the data readily available. As with any electronic data system it must be secure and safe at all times.

Data Security

Making a system that stores lots of personal data requires excellent security measures to be put in place.

Credit card companies expect all of their merchants to comply with Data Security Standards (DSS) as outlined by the Payment Card Industry (PCI). The PCI DSS clearly defines procedures for storing, transmitting and accepting credit card data.

Those found to be in breach of these procedures are swiftly dealt with.

Resorts too have an obligation to ensure a patient and witnesses' personal details are correctly handled. The DSS is an excellent source material to help set your own standards.

You can read more about PCI security standards at *https://www.pcisecuritystandards.org/*

When building your own IT systems, your development team should know and understand the risks surrounding data security and ensure they are complying with those industry standards. They should also have a defined plan of system updates to ensure that all levels of the system are up-to-date and secure.

User process

A software program is only as good as the person who uses it. A developer can put all the measures in place to secure the data from external hackers, but if a user enters information in the wrong place, or copies it into an email then the data becomes unsecure.

As obvious as this may sound, I have had to deal with companies who have been hacked, and unencrypted data has been lost. One company racked up just over $100,000 in fraudulent credit card losses. This is the threshold at which Visa and MasterCard take a very active interest in the issue.

In this case an ecommerce website was compromised, after an opening was spotted by hackers. The hackers were able to push in special codes to the web address that allowed them to return unencrypted credit card numbers. They were able to do so because the developer had not taken required precautions to prevent this type of attack.

Additionally, I discovered that company employees were not saving credit card details in the appropriate encrypted fields on the data collection pages. They were taking credit card numbers and putting it into the general call notes field, with expiry dates and CVC/CVVs. This meant any user could read and take the credit card information, and sell it outside the organisation.

As you can see it is crucial that all users are given full training, regular updates about data security measures and are audited. They must be made aware of the sensitivity of the data and take appropriate precautions to protect it.

Another way to secure data is through the design of the application. By setting security parameters you are able to limit visibility of sensitive data. That means users will only be able to see information that is essential to their job role. These actions can help you ensure you are in line with legislation such the *US Health Insurance Portability and Accountability Act, Title II (HIPAA)*.

For example, a company legal department would need full access to an accident record, but a manager involved in risk assessment, or a patroller, would not need to know the patient's home address after the fact. Likewise a patroller may need full access to another patroller's files for a short time during or after the incident management.

Portable device security

In a world where we now have 'bring your own device' (BYOD) culture in offices, users can bring their own smartphone for work rather than having a dedicated device provided by their company.

This means that company information is carted around with the employee on personal time to cafes, restaurants, gyms and even on holidays. This leaves the company with challenges around security of data.

So how do you manage all of these issues? First, you should involve your human resources department to produce agreements for staff who use the system, linking their responsibilities to their employment contracts.

Secondly consider a piece of security software to install on the users' smartphones that keeps the company data separate, backed up and if necessary gives your IT department the power to remove sensitive information from the device remotely. This can be used in cases where the phone is lost or stolen, and when the employee moves on.

Office and location security

Given the source of data is sometimes offsite or on a mobile device, you must also ensure that your offices are not vulnerable to attack from the outside. A physical office connected to the internet should be treated like a data centre. You should ensure that it does not allow unauthorized access from outside, and that appropriate measures are taken to protect the data flowing in and out.

Physical security measures to consider are: who can enter a building or office, and what measures are taken to stop unauthorized access; installation of cameras and logging of swipe cards use to understand "who went where" when an investigation is required.

You should also look at the security of data and ensure computer terminals have logins for each user, enforce regular changes of passwords, and store backups in a secure location, and duplicate them offsite.

Server security

My bet is that your current system of storing accident reports is a box, in a cupboard, that isn't fireproof. You probably don't have a copy of them either. Ask yourself what would happen if you did have a fire, or a theft, and one of those cases then ended up in court?

A server (like a web or database server) should be in a secure data centre, one that has no single point of failure. The means having at least two of everything including fire systems, internet connectivity, physical security access, power supply, backup systems, and hardware set up.

If you have never been into a data centre, imagine a large building within which sit long lines of computers, on many floors. To get in, you must enter through security, have an induction and go through multiple sets of specially sealed doors. It's a bit like trying to get into the Pentagon!

To get an idea on how this works, you can check out Google's data centre in a controlled manner on Google Street View.

http://www.google.com.au/about/datacenters/inside/streetview/

Data transport security

Making sure the data is safe between your devices, data centres and offices is one of the simpler parts to this puzzle. You must encrypt it as it passes over the public internet to ensure there is no snooping and copying of information. This can be done in a number of ways but the most convenient way is via the Secure Sockets Layer (SSL) certificate. This consists of a public and private key and allows devices that know the keys to encrypt and decrypt the data. These are commonly used on ecommerce websites where personal or payment data must change hands.

Backup

Your data should always be backed up, preferably in a second secure location. This is much easier now than it used to be when dealing with tapes (and taking the tapes offsite). We can now take advantage of cloud technology and make backups over the internet on a secure channel to another server in another data centre. For example, you could have your main systems in Dallas, USA but your backups are in Sydney, Australia. This is not hard to do and shipping the data does not take long to complete.

It's also very easy to create entire system backups (called images) of a server at a point in time. This can be very advantageous in the case of hardware loss or failure. These images take all of the system settings and data, which can be applied to a new set of identical hardware in a short period of time, cutting out configuration and set up time, should something go wrong.

Active monitoring

Finally, you should be taking steps to ensure that you are secure by using a third party system to scan your offices and servers on a daily basis. The PCI standards provide an array of services and systems that will conduct penetration testing and advise you where the weaknesses are in your system.

This proactive monitoring allows you to find and patch weaknesses in your systems before outsiders can exploit them.

A mobile phone can collect and keep this data accurately and safely from the start of the process through to handover to doctors or ambulance personnel involved in the chain of care. We will discuss this idea, and its implementation in more detail in the final chapter.

Workbook / Tips

- Can you find 50 accidents in the last season and accurately place them within 3 metres of the actual incident, on a map?

- List the preventative actions that occurred in this area, and see if they made a difference or simply moved the issue to another location.
- Write down your medical handover procedure, and list the information that is passed along with every patient.

Technology

Technology is fast becoming a major asset for ski resort operators – you already use it extensively in many areas, but when it comes to risk management, data collection and incident reporting of skiing accidents there is a serious gap in the take up of technology.

Using technology, which is readily and cheaply available, now provides us with the ability to manage risk and data collection more than we ever believed possible.

Great technology is never seen. People use it and really only notice it when it doesn't work or when they can't achieve their objectives. New systems must be designed to fit in with existing workflows and be easy to use. Where possible, it should be designed to actually save the user time, and allow them to be more effective in their job.

We see improved technology in the form of more efficient, environmentally friendly and reliable lifts; better ticketing systems to track users; and reduction of queues for ski lifts through buying tickets online.

So why don't we see the same level of technology available to ski patrols or the medical care teams for management of their accident data, and communications?

It is certainly available in the form of medical equipment used to save lives as seen in the use of portable defibrillators and the adoption of the latest research on CPR and patient care.

However I am not talking about medical equipment. I am referring to technology that improves process, communications and management. We still don't see the influx of technology on the slopes that is available to those in other industries.

Those further along in the chain of care, such as paramedics, medical centres and hospitals have access to better systems for data recording, communications and analysis but some of the most important information is from the scene in the first few minutes after the incident.

For a lot of resorts, this lack of technology is down to money and time. The larger resorts and management companies have the ability to develop these systems, however they often focus on product to enhance the consumer experience. Vail resorts has, since 2010, been developing the EpicMix system for its guests to track their vertical descent and number of lifts taken. This system is using an industry technique called gamifying. Their app tracks the user's day, encourages them to complete missions and achieve rewards that they can share with others. Users can purchase photos taken by on-snow photographers, and share the day's events on Facebook and Twitter. This link to social media spreads the resort brand.

You would think that with all the advancements and investment in the industry we would be on the cutting edge of technology, ensuring that patrols not only use the best equipment but that as life savers were doing everything possible to ensure the best care for our patients.

I think that in most cases, there is a lack of understanding of what is possible through the use of new technology.

In early 2013 it was estimated that 50% of the adult population of the USA use smartphones and regularly use mobile apps. By 2015, that will grow to be 80% of the adult population. That's a lot of people using this 'new' technology.

Globally the uptake of smartphones and apps is just as significant.

The spread of smartphones is faster than any other technology before it. It has been so fast that we have yet to discover its full, applied potential and to scratch the surface of what can be done. After all the majority of apps available today are games.

In the 'old days' we would create a product to solve a particular problem in a particular industry. Now we have a device that can be applied in so many ways, across so many industries, that we are all looking for ways to take advantage of it.

Based on experience and imagination a user will often have an idea, and they think about what the device could do for them. Those thoughts and conversations usually go along the lines of "Wouldn't it be cool if…?"

We are the most technologically advanced beings on the planet. You may have years of experience in your job, but I bet your children can operate an iPad better than you. Someone said to me recently that they don't think their six-month-old twin boys will ever learn to use a mouse with a computer - they will only ever use touch screens. I agree, because the younger generation is already growing up with technology in every aspect of their lives.

If you don't take on this technological change, it will ultimately result in your brand being weakened and your business will suffer.

Lucky for us, smartphone technology provides an inexpensive way to improve this process.

Apps like Perisher Blue's or Vail's EpicMix are currently used for marketing purposes and are putting this technology into the customers' pockets already. It would be easy to provide the user with a way to report their location to the appropriate department in the resort through those apps. Granted it may be within 20 metres or so of the actual location, but it's better than needing to send four patrollers out on a hunt. Let's have a look at some of the systems available to help people in distress get help.

International distress systems

Emergency Position-Indicating Radio Beacons (EPIRB) are designed to get the call out from patients faster from remote locations.

EPIRBs are battery powered, water proof devices that report their position via an international satellite network and are commonly used on marine vessels.

In the Blue Mountains of New South Wales, mobile signal is patchy in rural areas. Even with Australia's best mobile network, you don't have to go far from civilisation before the landscape gets in the way and you simply have no contact with the outside world.

This often results in unprepared adventurers getting lost. In recent years, a number of high profile media cases have highlighted the difficulties in locating people.

To help combat this problem, the local rescue associations and police have come together to provide Personal Locator Beacons (PLBs are mini EPIRBs) to people who feel they need a safety net. Unfortunately they are not mandatory.

These small, handheld devices work in the same way as a marine or aircraft EPIRB and transmit a signal to low orbit satellites. The signal is forwarded on to a land-based centre that identifies the country of origin and alerts the relevant country coordination centre. This is handed down to the authority covering the location of the beacon and they initiate appropriate search and rescue protocols.

The beacon also sends out radio signals on international search and rescue distress frequencies for the rescue team to monitor. The beacon can be tracked in this manner too, although it's harder to do so.

In February of 2009, the beacon system changed to a digital one, and each beacon is now individual, so it can share not only your position but your personal details too.

These systems are impractical on a large scale such as in a ski resort, but are great for those venturing outside the resort boundaries and into the wilderness in winter.

'Mobile' people trackers

There are a number of apps now available on smart phones (iPhone/ BlackBerry/Android) that allow a user to track their day using the

GPS component on their phone. These apps poll the phone to find out the location and altitude on a regular basis and by comparing data points can discover the user's position. Speed, G-force and descent or ascent speed can also be calculated.

The data can also be put on a map and displayed visually.

Lots of these are designed for outdoor activities but there is only a handful specifically for skiers and snowboarders.

Smartphones

Market indicators in 2012 show that more than 50% of the population uses a smartphone. This is a trend that follows in the US, Australia and the UK. The growth is so fast that market saturation will be reached sometime in 2015. This means that in the US, 225 million people will own a smartphone out of the 310 million population.

Resorts are keying in slowly to the power of the mobile device and providing engagement apps to share their photos and day's activity across social media.

Going back to the customer experience, these apps are not only adding value to their day, but they help spread the resort's brand. The prevalence of these devices also makes them a low-cost and powerful way of connecting people and information together in a widely spread geographical space.

Let's look at a couple of examples on the market right now:

EpicMix
Vail Resort's free EpicMix ski app "allows you to track your vertical feet, connect with friends, and share photos from your ski vacation".

Although EpicMix doesn't provide exact run paths it is connected to your lift pass, and works in conjunction with the resort ticketing system. It can report on how many runs you make and which lifts you took. This translates into vertical descent and runs taken in a day. With the social interactions such as checking in and linking to friends, users can earn awards along the way. These are small wins, which are acknowledged by the system to keep people engaged.

The Vail resorts also send professional photographers around the slopes to take photos of guests on the snow. These photos are linked to the guest's pass and sends the photos through to the app. This allows the resort to create a better return on the photos as the guest can then view them on the phone and purchase copies.

Guests who use the application and share the experience over social media are providing brand spread by getting involved in the online as well as the physical worlds.

EpicMix is built with marketing and brand reinforcement in mind. It currently has no ability to report the user's location with a request for help. The user still needs to make a phone call to the ski patrol, and describe their location.

SkiTracks

SkiTracks is a standalone iPhone application that knows where you are at any point in your day.

This application tracks your day using the smartphones built in location aware functions like GPS. It constantly measures where you are and calculates your speed, altitude and more. At the end of the day you can see how many runs you made, look at a map of your day and share it with friends online.

Applications like this are making great use of the technology available to us in smartphones however they use a lot of power. On days when I have run this application, my phone's battery charge has only lasted until lunchtime. No good when you are not near a power point and you are out in the resort from first to last lifts without a break!

Because this application is independent to a resort, it also has no way of reporting the user's location in order to get help.

Some background about mobile GPS

Most smartphones don't rely on GPS data for location because it's expensive in terms of processing and battery life. Instead, a phone will use its cell towers and Wi-Fi networks for triangulation. Where only one or two cell towers exist and there is no Wi-Fi, the phone must activate the GPS chip.

You can see how this works by turning off your phone's Wi-Fi and then opening up an app that needs to find your location. It will prompt you to turn it on for a faster, more accurate location.

The problem with GPS is the antenna size in a smartphone. Your phone is trying to listen to a signal from a satellite at least 20,000 kilometres away. This is like listening for a mouse squeak from an aircraft through all the background noise. Amazingly it is possible!

Currently there are 31 GPS satellites orbiting the earth for public use. One is for testing, six are spare and the remaining 24 are in full use. They are positioned in such a way that you can see up to 12 at any one time. The more satellites your device can connect to, the better the information and calculation of the position, so the result is more accurate.

Most modern smartphones now support GNSS (Global Navigation Satellite System) which means it can work with both US GPS and the Russian managed GLONASS satellites. That means you have a better constellation of satellites to draw from when you are in the field.

Dedicated GPS devices with a decent power rating and antenna will be able to calculate a position down to about 300mm. Where other GPS sources are available, like the land based Wide Area Augmentation System (WAAS) this increases the number of GPS sources and can help increase accuracy again.

To help get a point of fix faster, there is a system called Assisted GPS (AGPS). This system provides the device with a simple file of data containing a map of where the satellites are in the sky at that moment. Instead of waiting for a signal from a satellite, the device can pinpoint it using the AGPS.

Android devices and iPhones augment the GPS signal when it is low or unavailable by using crowd sourced Wi-Fi location information and cell tower triangulation. This is clearly not so useful in the backcountry or the South Pole.

How your phone knows where you are

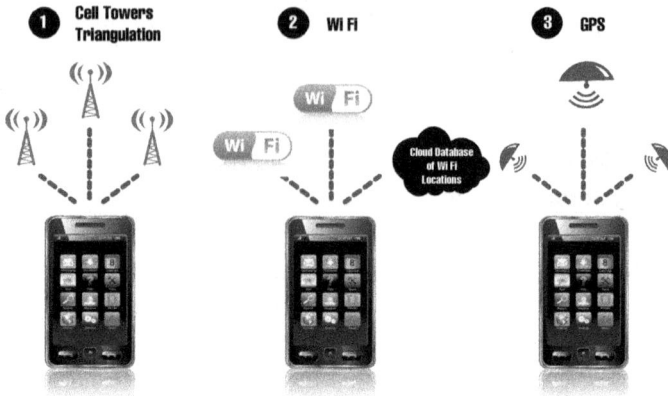

So what are the realities here? Typically a phone using only GPS will find your point within about three metres in open skies[35]. That's pretty good, and is more than accurate enough for the purposes of marking accidents on a resort map.

We have all experienced reception issues at some point due to our location. However you would be surprised at some of the places where cell phone reception is available. For example on Mount Kosciusko, Australia's highest point at 2,228 metres you can have a chat on your phone with your grandmother in the UK.

Most ski resorts will have reasonably good coverage because they are populated places. Networks are continually responding to the demand for increased coverage.

By using a smartphone for reporting accidents, we can collect a location for each.

This helps to cut down response time, and provide useful analytical data for reporting. As technology improves we will get more accurate and closer than the current 30 metre radius.

High power RFID tags

Radio Frequency Identification (RFID) tags are commonly used in security passes, lift passes and other contactless card systems.

These use very low power and work on a short range. A high power version of this system is available for use over larger areas.

Because Ultra High Frequency (UHF) signals work like a fountain spray projecting up and then falling around you, it can get around terrain that cell signals cannot reach. The technology this uses may be familiar to you. It is an RFID tag. This works by transmitting an encoded number on a certain frequency to anything close by that will listen. In fact your modern day hands-free lift pass uses exactly this system on a low power high frequency band. This projects your pass number in a straight line from your pocket, and into the receivers at the lift terminal. RFID tags are also used extensively in retail for theft protection, in your passport for identity and warehouses for stock inventory.

Imagine that your ski resort has a series of receptors on each lift tower and buildings around the resort. If three of these receptors pick up a radio signal, they can be matched together to calculate the location of the transmission.

The transmitter would be no larger than an Oreo cookie, easy to fit in your pocket, and at the right scale would cost about $5 each. They would only transmit if the user pressed the button, much like an EPIRB, but instead of going through the process of alerting national organisations down through a chain back to the patrol, this is a local system.

Installing the reception devices wouldn't be hard as it can be hooked into the existing power on most lift towers.

A system like this in a ski resort could cut down the time required to get help to a guest, and could even work outside the resort boundary with the appropriate reception points.

SMS Beacons

Short Message Service (SMS) beacons are tiny devices that can send text messages with a piece of information to a pre-programmed number over cell phone networks.

This system has been coupled with a GPS receiver to create a location aware SMS Beacon. These have a retail cost of between

$50 and $100 and of course require a pay–as-you-go sim card. The advantage over a full phone is that the battery life will be significantly longer, and it only needs a small cell signal to get an SMS out. These devices are popular with hikers as a backup device.

Avalanche Beacons

Avalanche beacons are personal devices worn by mountaineers, skiers and snowboarders who adventure in avalanche-prone terrain. They perform two functions: transmitting a signal, and as a tracking device to find another's signal.

These are worth a mention only to acknowledge their place in finding a patient. Due to the short range of the transmission of an Avalanche Beacon, any rescuer needs a fair idea of where the patient was last seen and where the path of the avalanche will have taken them. Any attempt to rescue with a positive outcome needs to occur within a few minutes. This is not a mass market solution to finding patients faster in a large area.

Patients using a smart phone

Creating an application to assist in reporting accidents is a reasonably small task and can make a big difference in getting the word to the patrol and cutting down response time. A member of the public could report it from a chairlift as they pass, or standing on scene with the patient. Their report could also include their phone number in cases where follow up or clarification may be required.

Smartphones are the most advanced piece of consumer technology available today. They are low cost and have become widespread in a very short time. No longer are we using these devices just to play games and make phone calls, we are now seeing applications that use the camera, voice technology, storage capacity, and interactivity that they provide.

This presents an amazing opportunity to resorts to change the way information is collected and communicated. A smartphone can be used on the ground to get timely and accurate information that can be subsequently analysed.

Furthermore, the support for smartphone accessories has grown and niche markets have appeared to provide products to those in unusual circumstances.

For example, there are attachable devices to measure many things such as heart rate, blood pressure, or blood sugar level. These devices provide even more scope to understand what is happening to a patient, where previously we would have needed expensive and bulky equipment.

By using a smartphone for reporting accidents, we can collect a location for each.

Workbook / Tips

- If your resort has an app, how many installs does it have, and how many users are active on it?
- How do you currently get alerted to a patient in the resort?
- What is your time to get to a patient from the time they had the accident? Could this be improved by technology?

Transition Management and Training

Introducing new ideas and work procedures to a business is challenging. By understanding how to effectively train your staff and manage change you will have a superior result.

Smartphones are the future. Introducing smartphone technology into your resort is inevitable. Have you noticed how many resorts now have an app for customer engagement? Despite their popularity smartphones are still a major challenge for many people, which is why planning your implementation is so important for mostly low-tech ski patrollers.

The key to successful change is understanding the difference between training, development and education.

Training

Training is concerned with learning the necessary skills to conduct job-related tasks. In the workplace, training allows teams to undertake their tasks correctly and effectively, based on the documentation you provide explaining the procedures and rules.

Education

Education relates to the broader activity of learning concerned with the holistic growth of the individual. Education allows trainees to understand why they are being trained to do things in a particular way.

Development

Development implies the growth of an individual in a non-organisational context as well as in the workplace. Development is mostly external to your organisation and allows the trainee to understand the greater implications of education or to apply other personal learning from practical experience and social feedback.

In this chapter we will be concentrating on the training aspect, and touching on education when looking at how to implement a new way of working.

Understanding how ideas and innovations spread

Before we jump in and look at how to initiate change I want to share a couple of ideas with you that can support you in managing expectations to give you the best chance of success.

The Diffusion of Innovations theory

The 'Diffusion of Innovations' theory explains:

"How, why, and at what rate new ideas and technology spread through cultures."

In broad terms this theory categorises people according to their decision making process, specifically how quickly they take on new ideas.

Understanding how people accept new ideas and products will help you implement change in your organisation.

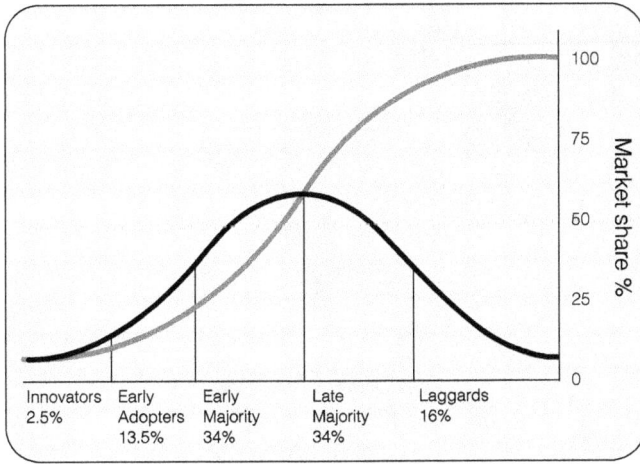

Chart showing market share % adoption curve with categories:
Innovators 2.5% | Early Adopters 13.5% | Early Majority 34% | Late Majority 34% | Laggards 16%

Innovators

Innovators are the people who find new ways of doing things. They create the concept, promote the idea and work out how the innovation fits in with their life – helping to define the benefits and possibilities.

Outside of a company, innovators are usually the youngest, or they have sufficient financial resources to cope with an idea that might fail. Both groups have the largest tolerance to change and risk.

Within companies, innovators are the people who push for the new way, or go out and develop the idea and make it happen regardless of the company position.

So who is this person in a Ski Patrol? Well, usually they're the one that has a stream of little projects on the go designed to improve your working day. Viewed as a maverick by others, the innovator is always looking for things to tweak.

The Early Adopters watch this person with interest, and help develop the ideas they see value in.

Early Adopters

This group is open minded, and very passionate about the area in which the innovation is occurring. Early adopters are influential in their peer groups and are well regarded for their opinion. However early adopters don't normally innovate as they are more discreet about the risk they choose to take on.

In a ski patrol environment early adopters will buy new equipment such as skis and avalanche beacons as soon as they become available. They like to understand the technology and work out how it can improve their lives.

They see value in a new piece of equipment for their own benefit, or that of others, and like to understand all options. They will help inform new policy and procedure within the patrol using their range of knowledge on the subject.

Early adopters are passionate about the things they love and will volunteer to help teach others about it, in particular the Early Majority.

Early Majority

This group is considered to be the bulk of people you require to have a system work and be sustainable on its own. Their adoption of an idea or process will come from contact with the Early Adopters who are already convinced of the merits of a product.

Within a ski patrol, the Early Majority will wait for things to become the rule before they adopt. Through training they will learn about the procedures and work it into their daily routine. They bring about alterations and adjustments as their experience improves providing feedback as they go.

Late Majority

These people approach new products and innovation with a high level of scepticism and do not have enough knowledge about the innovation to see that it is a benefit to them. Because they are not open to change their learning and rate of adoption is slow.

The Late Majority will only adopt a new product or procedure because a) they have to, or b) they are 100% convinced of the benefits. Even after adopting the product the Late Majority will need a regular push to ensure they don't slip back into old ways.

You will find almost continuous resistance from the more stubborn members of this group, those who are on the border of becoming Laggards.

Laggards

These are the people who almost actively resist change and are focused on "traditions". They tend to be the oldest group and have little or no contact with the Innovators and Early Adopters. They are usually the smallest group.

Laggards usually require an 'official stamp' of mass approval before they will take it on.

Conscious Competence

This theory looks at the four stages of competency people go through when learning a new skill.

People grow into skills through a number of levels of competency. Consider driving a car.

As a beginner, you have spent 16 years watching Mum and Dad drive, so you think you can drive too (unconscious incompetence).

You get in to the driver's seat, and realise you can't (conscious incompetence).

You're going to have to spend lots of time doing this and that, learning and checking everything. Soon, you can drive, but you have to concentrate hard (conscious competence).

After doing it for some time, you don't even think about it (unconscious competence). You just drive. It's automatic so you can talk, listen to music and watch the scenery go by as you drive.

Unconscious Incompetent

People in the first stage essentially don't know what they don't know. This is because they have no need to be able to complete a task or see no value in it, and so put no thought into the required skill. However when they do think about it, they believe they can do it but have yet to discover they can't, and will require lots of practice.

To move beyond this stage, a person must have a willingness to learn and recognise the benefit of the skill.

Conscious Incompetent

As a person tries something for the first time, they discover they are lacking necessary skills to complete the task. They start to understand the areas in which they need more practice, by making mistakes as they go. This is integral to learning the new skill and progressing.

Conscious Competence

People start to feel competent when they are completing the new skill to a level they feel matches their expectation of quality. But in order to do it, they must concentrate hard, and have very little ability to do anything but that task. With practice this becomes easier.

Unconscious Competence

Finally as the person becomes very skilled they are doing things without realising it, or thinking about the minor details. It's all just happening and it's second nature, like muscle memory. As a result these people are able to take on another task simultaneously.

There is scholarly debate going on to decide if there is a fifth level, which is missing from this model – "Conscious competence of unconscious competence".

This is the stage where you have the ability to develop unconscious competence in others, that is, train others to the level where they can multi-task. It's really hard to teach someone else without being taught how to teach, or being conscious of the mechanics. You will struggle to unbundle the automatic routine and break it down to the component parts needed to develop through the stages. Indeed the best at the task, the unconscious competents, are the worst teachers because they cannot break down the task into its parts to be able to educate others.

Take this in to account when you train people on the new systems. Allow them time based on age and learning style to become familiar with the new system and take it on board.

Initiating the change

When you are the leader or innovator, change starts with you. No-one else is going to push this forward, and you must champion the idea relentlessly until you reach your goal. So let's create a plan that will help you implement your idea.

Look at how you can make a change to have your resort move from paper based accident reporting to digital methods.

Pre-planning

The key to the successful implementation of any change is being clear on what it is you want to change and why.

Having an action plan covering the wheres, whys, and hows is crucial because:

- It will help you clarify your thinking.
- It helps you spread a consistent message.
- You are creating an asset from which to build your procedure and training material.

Questions to think about include:

- What are the benefits of your idea to the business, your resort and guests?
- How much will it cost to implement?
- Will there be a saving in the longer term?
- How long will it take to implement?
- Does it break another system while you are introducing it?
- What happens if it doesn't work?
- What are the legal implications, if any?
- Is it practical?
- How will it affect your staff and team?

After answering these questions for your resort situation, the next step is to check your ideas with senior members of your team. Share your idea with a mix of patrollers consisting of early adopters, trainers and others with experience to get a balanced opinion. The outcome of this meeting will be to get support, and refine your concepts.

If you need funding or formal support from management this is the time to take it upstairs.

Setting up the rollout

With backing from your trusted patrollers and support from management, now is the time to work out how to deliver it for widespread adoption.

Questions to consider include:

- Do you need a prototype or focus group to refine the idea?
- Who are your learning groups, and in which order should they be trained?
- Can you train on the job, or do you need classroom, reading or presentations too?
- Do you need to adapt existing training manuals, or procedure?
- Does this affect staff contracts or legal matters?
- Do you need to purchase any additional equipment?
- How much time will it take to train?
- How would you measure a participant's level of understanding, and the success of the training?

Answering these questions and working with your training team will help set you on the correct path.

Measurement

Change should always be measured for its success. How else are you going to know if your implementation is worthwhile?

Questions to consider include:

- Did the project run on time?
- Did the project run within budget?
- Has the project saved money, or created new or repeat business? If so, how much?
- Did the project increase or decrease the time required to complete a job?
- Are your guests making more runs in a day,

or spending more money?

- What else could impact these measurements?
- Your measurements need to be based on fact rather than opinion.

When you know the answer to these questions, take your measurements now, and again at regular intervals as you roll out your training. Ensure you are aware of the other factors that influence these numbers, and if possible refine the measurement so it is as focused as possible.

Training the early adopters

Once your plan is in place and you know how you will monitor the change for success, you are ready to get on with spreading the word.

Train your first group, probably your trainers and a couple of early adopters. The trainers need some time to absorb the procedure. As they are in a position of authority, they will be comfortable in providing feedback to you. They will also help tweak the delivery and training method early on as they are being trained.

Change your plan as necessary and then have your trainers roll out to the rest of the team on the schedule you defined.

Also consider members of this early group to become leaders of the program.

Help them do this by rewarding them with a job title, decision-making power, and maybe a financially related incentive. This will encourage them to step up, be engaged and encourage others to follow.

Review the training

Once your timeline has progressed and training is in full swing, take the opportunity to sit back and admire your work. Don't stay there for too long because you need to know how successful this has become.

Go back to your measurements and talk to the trainers about their ability to deliver, and the response they received while teaching the topic.

Questions to ask:
- Was it easy to deliver?
- What were the common sticking points with each group?
- Who had issues with the concepts, and why?

Monitor your newly skilled team and find out how they liked the training, and how the new procedure is fitting in to their normal workday. You are looking for ways to improve the training, the execution of the procedure and to find other issues which you can rinse out and fix up for the next round of training.

Glenn Kirkwood, the training officer at Thredbo, reinforces that you need to know how to test people on the training, and ongoing competency. If you don't, then how do you know the results?

"You've got to have a base line to assess your change. You can't have a massive change and then not measure the results, otherwise you won't know what worked and what didn't work. So in the second round [of training], you might only make one or two changes based on the first season of training. By the second season you're saying: "We did this and that, it went better or it went worse. Okay well, when it went worse, take that out and put something else in. Did that go better or worse, okay that went better. Okay we'll leave that in, let's try something else. Better or worse?'"

This continuous evaluation and change allows the training to be more efficient and effective, making it not only easier to teach but easier to become proficient. It also maintains a fresh look at the system and keeps your procedure up to date.

CASE STUDY 4: THREDBO SKI PATROL

In Thredbo, the training system is an integrated approach to pulling people through the ranks with a good blend of practical learning on the hill and structured learning with an instructor.

Application

Trainees start with an application to the patrol to enter the ski test. This test happens once a season and is designed to help the candidate show their core skills. These are: to ski crud on a poor conditions day, snow plough, side slipping, and general ski technique. Each discipline is demonstrated by a senior patroller so the candidates get the idea and can watch what is required.

First Aid Training

Independently of what they do on snow in the winter season, the trainee must attend a classroom first aid course to qualify in *Emergency Outdoor Care,* or an *Emergency Medical Technician* course. These advanced first aid courses teach the essential medical skills to care for a range of injuries in a ski resort.

The principles can be used for many other situations, such as administering first aid if you were to be the first to a car crash. Candidates learn how to use equipment like scoops, backboards, splints, defibrillators, provide CPR and administer pain relief.

Most countries outside of Australia require patrollers to do an avalanche awareness course to understand the signs and risks of avalanches. These courses also go into the basics of cutting cornices and other avalanche control techniques and preparing the trainee for some of the conditions they may come across.

Their first season is made up of work with patrollers on days they are on the hill. Their classroom learning is set up as four, two-day sessions with the training officer, and one full week. The end of the week has the exam date, so everything in the week prior is geared to ensuring they are ready for the test. The training focuses on on-hill skills such as trail work technique (knots, fencing, signage, and sweeps), radio communications, mountain knowledge, daily procedures, medical centre procedure, medical scenarios and sled running.

The Trainee Test

The test is comprised of two parts. First the knowledge test proves the candidate has learnt about equipment, personnel and the mountain itself. Questions are asked about sweeps, radio call signs, trail names and access routes. The second part of the test requires the candidate to perform a demonstration of sled running technique with a test patient on board. This is conducted on one of the most challenging runs in the resort, ensuring they have the ability to conduct a rescue in any other location safely.

Passing the test, which can sometimes take two seasons for a volunteer, affords the candidate the coveted ski patrol vest and makes them a probationary patroller.

Probationary

At this point they are free to take their own steps in a resort and become one of the patrol team. The first job for a probationary is to spend a day in the busiest part of the resort helping on the beginners' area.

This area is covered by two patrollers with a snowmobile and an ambulance for transport to the medical centre. This area also acts as a transport point for sleds coming down on one side of the mountain. Because of the number of people in the beginners' area and the level of skill, injuries are common here and a good Saturday sees a lot of work by this team.

The probationer will be allocated to boost that team on a Saturday and gain fast access to many real injuries, lots of protocol and practical experience, while also providing a practical advantage to the usual two man team.

During the probationary period, the patroller is assigned a mentor. This person will usually be in the Nationals program, working to gain the highest Australian qualification available. The mentor is able to provide a sounding board to answer any questions, advice and to check in on their progress. They will act as a conduit back to the patrol management providing another view on the patroller prior to their completion of probation.

Feedback is gathered from area leaders and patrol supervisors on the probationer's input, attitude and work rate. Their last act as a probationer is to have a sled assessment to ensure they have not developed any bad habits and they are then approved to become a full patroller.

Somewhere in that process the patroller must take the huge step to be the first responder to an incident. This is a nervous time for anyone who must interact with a member of the public and potentially help save their life.

Nationals

In Australia, the Nationals qualification is the highest level of patrolling available.

The National patrol examiners assess candidates on ski technique and sled running in challenging conditions. This is a test usually undertaken with one to two seasons of training prior and lots of time on skis. Patrollers who take it will be experienced senior members of their resort patrol, having worked three or more seasons. It is designed to bring up the level of skill and knowledge of a patroller, which they can take back to their patrol. The test may be taken in another resort depending on the year. The Nationals system is used as the access point to the training team in the patroller's resort. This level of qualification sets the bar for quality and helps improve Australian patrolling as a whole.

Recertification

Every year, each member of the patrol is required to recertify certain medical procedures such as CPR and defibrillator use. Every three years, they return to complete a full Emergency First Aid course and must demonstrate good technique and safe sled running on snow.

Benefit

The training system in place in Thredbo will be very similar to others around the world. It ensures that the would-be patroller is selected with the right base set of skills in skiing, which are then developed through the specialised skills of first aid and sled running and general patrolling. Enough time is put into each candidate to produce an excellent result and feed in great talent to the core of the patrol. As new blood comes in, the existing body of patrollers are provided with ongoing assessment and options to get involved in other areas of training. The system delivers a level of quality in the patrollers, which allows many of them to work in Europe, Japan and North America in their summer.

Workbook / Tips

- The key to implementing change is having a great plan, and the ability to measure its success. Understanding the people you are working with and how they adapt to change will help your training go smoothly. Who are the Laggards, Innovators and Early Adopters in your team? See the section "Setting up the rollout above".
- Who would you call upon to develop and initiate a training program?
- What resources would you need to undertake a training program, taking into account the range of staff involved, time constraints and budgets?

- Would a train the trainer program work?
- How would you measure success of any training program?

About Medic52

Prevention of any potential accident relies on a body of information. Medic52 streamlines the collection of data from every accident, building the amount of data you have faster than ever before.

Time and information are critical in the survival and full recovery of any patient, in any medical situation.

To prevent an increase in accidents action must be taken, through avenues like on snow work, education and insurances. Knowing what to do in these areas takes significant amounts of knowledge, and this can only be gathered by having access to data or full time on the groundwork.

Medic52 helps address these three problems by creating a system where communication of patient data is easily collected, privately handled, and made accessible for reporting.

Where did the idea come from?

After 10 years of patrolling in Thredbo, the patrolling group started to see the importance of using technology to overcome the issues we all experience with completing paper-based forms. Every patroller has experienced the painful process of having to re-key a number of fields into a Microsoft Access database on a laptop.

Because only a small proportion of information is entered in to the database most of it becomes useless as soon as the page is turned. Further to that, the database is used to provide only a small handful of statistics.

The loss of useful information at each stage is staggering.

What if a patroller could collect the information on scene? This would increase the coverage of more detailed data with photos, location and witness accounts, and mean that accident investigations would see a boost in the data that was collected first hand.

If we were to do this, then it could all go into a central database immediately the information is entered, making it easier to report on and identify interesting and useful trends from the data.

The final obvious extension to this idea was that patients are notorious for not knowing exactly where they are. The proliferation of smartphones means we can have the patient help themselves, and use the GPS to alert the patrol directly when help is required. This can cut down the resources required to find a patient, and the time required to do it.

What next?

It took me about two seasons to do anything about it.

After a number of conversations with my colleagues and a few IT experts, I decided it would be worth a shot. I finally had some space in my life after exiting my business at the time.

After a few meetings with Thredbo management, and gaining their consent to give it a go, I set to work.

I built a prototype in a piece of software similar to the existing database, however it is smartphone enabled, and located in the cloud. This meant a number of people could access it, and add data. We essentially ported the accident report form directly into the new database, and added some easy workflow to guide the user.

The purpose of the prototype was to understand how a patroller could and would use it. As the primary carers and data collectors they are the most important part of this puzzle, so it is essential that that they will be comfortable to use it. If not, then it's going to be a waste of time.

This microscopic analysis of the accident form (that has been developed over many years) turned over a few things that were outdated and no longer required. These were removed and I added in a number of elements that would normally cost a good deal to add to the paper forms, with redesign and printing costs.

Medic52 for patrollers

| ●●●○○ Optus 🛜 | 6:32 pm | 🔋 |

| ☰ | **Patient** | Next > |

| Male 🚹 | Female 🚺 |

CUSTOMER TYPE

Type   Guest >

PERSONAL DETAILS

Full Name

Date of Birth

Email

CONTACT DETAILS

Address (Line 1)

Address (Line 1)

Suburb

Entering Patient Details

History   Next ›

Activity   Snowboarding ›

Ability   Beginner ›

EQUIPMENT DETAILS

Equipment   Rented ›

Rental Shop   Name and address

Wrist Guards

Helmet

Body Armour

SKI DIN SETTINGS

Front   Left   Right

Back   Left   Right

MEDICAL HISTORY

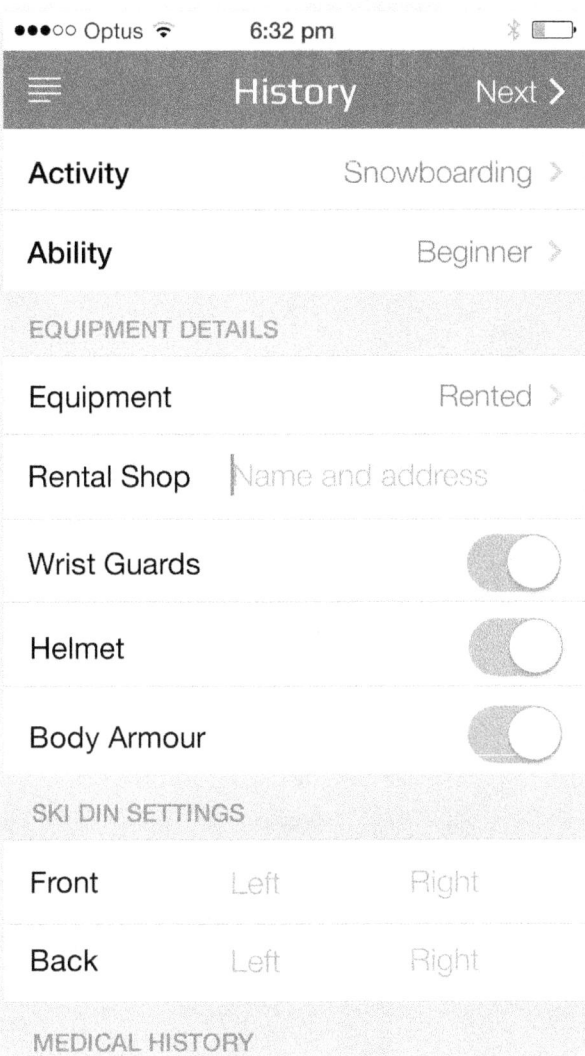

Collecting Patient medical history

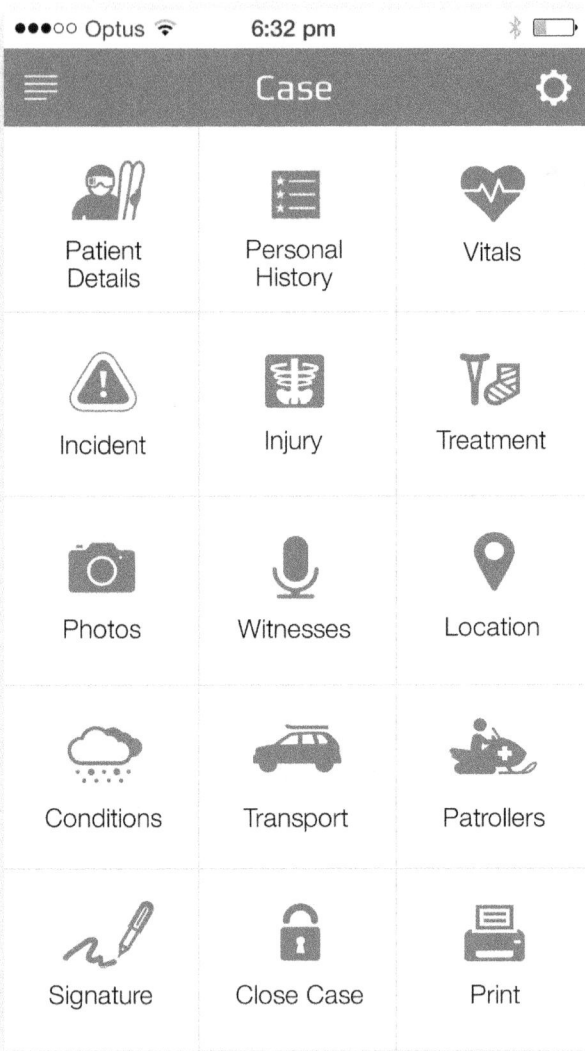

The Case dashboard, providing quick access to all the parts of the incident report.

Entering an injury for the patient, multiple injuries can be recorded for different body parts

☰   Take Vitals   Next ⟩

PULSE & BREATHING

Pulse   BPM

Breathing   RPM

GCS

Eyes: 4 / 4   Open eyes to command

Verbal: 4 / 5   Open eyes to command

Motor: 5 / 6   Open eyes to command

Total: **15** ⚠

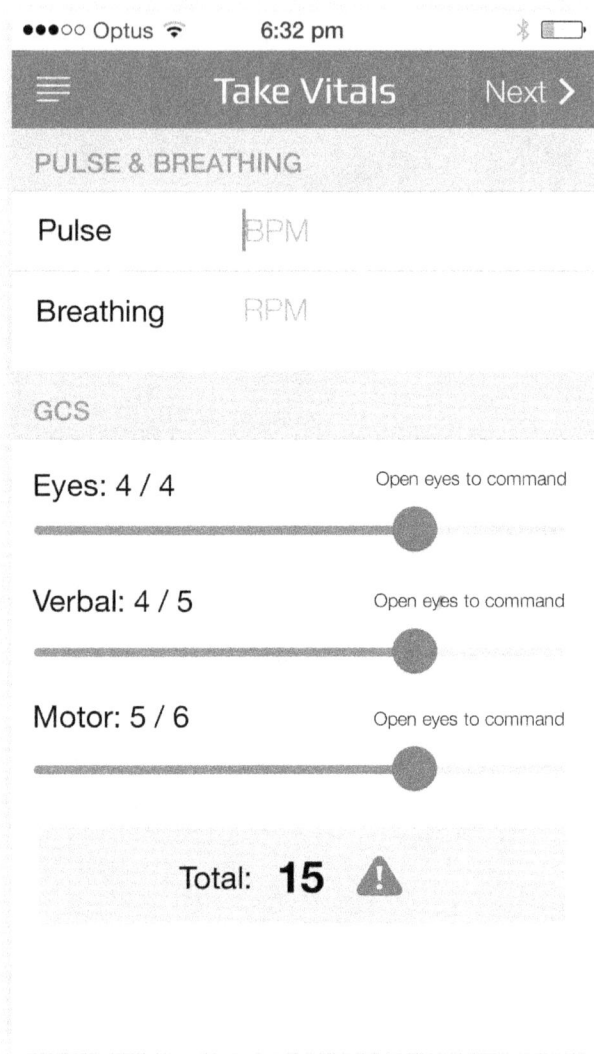

Collecting Vital signs using the Glasgow Coma Scale, and collecting pulse and breathing rates to build up a valuable picture of the patient's situation.

≡   **Vitals**   Next >

		220					15
176							13
132							9
88							6
44							3
0							0
		1		2		3	

GCS	BPM	RPM		
15	80	30	7 mins ago	❗
15	78	28	14 mins ago	◗
15	85	26	20 mins ago	◖

＋

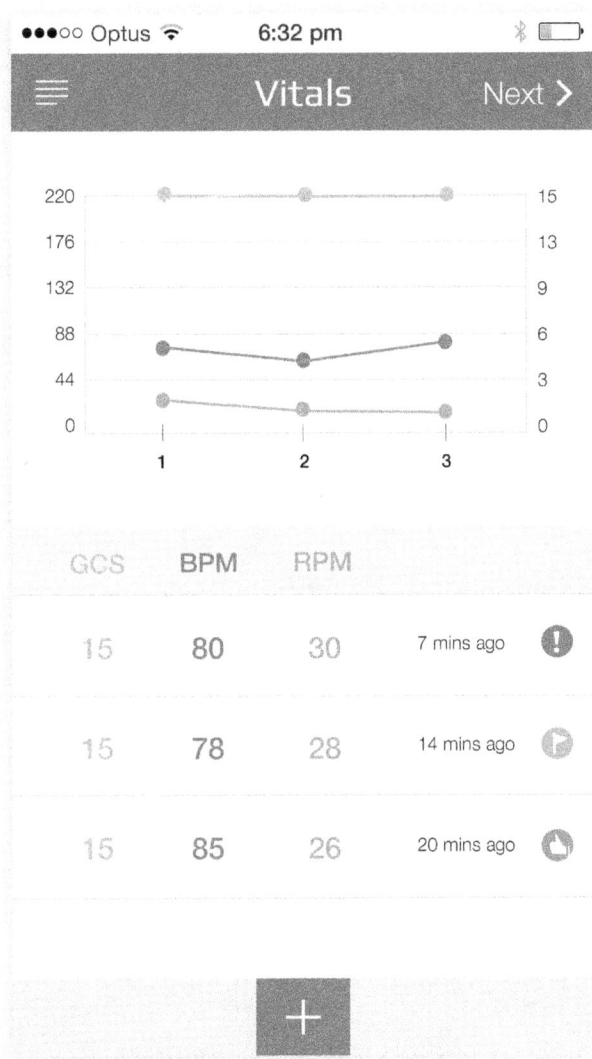

Vital information collected over time shows an overview of the patient status.

Body Parts Save

Head

Neck ✓

Face

Shoulder

Clavicle

Face

Nose

Teeth

Upper Arm

Lower Arm

Hand

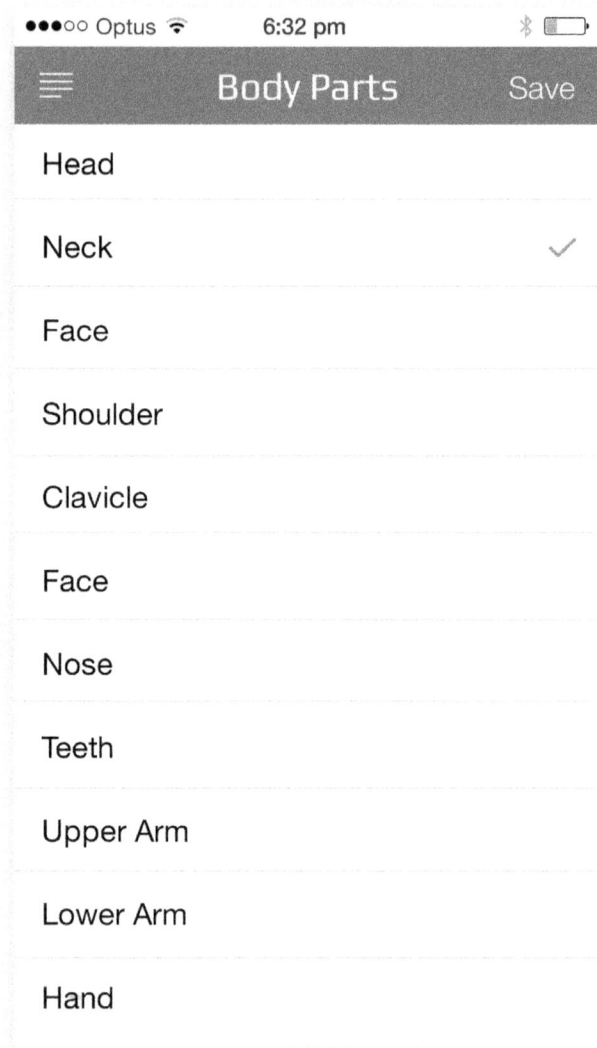

Selecting a body part or injury is easy from the list.

We had some patrollers trial the app, and discussed the merits and procedures by which it would be used. The feedback from the patrollers was that they greatly appreciated that the data does not have to be double handled any more, and that a lot of the 'paper-work' can be done on snow. This means less time in the medical centre and they are back on hill faster.

Here is some of the feedback we received after the initial testing:

"Data collection efficient & accurate"
- Jo, Thredbo Patroller

"Accurate, efficient real time data collection"
– Ben, Fernie & Thredbo Patroller

"Overcomes problems with illegible handwriting"
– Simon, Supervisor and Val D'Isere Patroller

"It will take us to the 21st Century"
–Adam, Niseko Patroller

In November, 2013 I attended the London International Ski & Travel Exchange (LISTEX) to present the ideas and gauge the reaction of a more international community.

The presentation was pitched to answer a question that hit home with the audience "Are accidents bad for business?" The reviews were strong with many taking the information back to their home resorts. We have since been featured in *The Ski Trade*, the UK ski trade magazine.

What about cell phone signal?

This is the most obvious challenge of using a smartphone. Many people have issues with signal in the middle of a city, so how would it work in a ski resort?

Most coverage is driven by demand, so telcos will go after the densest regions first. A ski resort is densely populated for a short period of time, and so there is interest to have coverage, but not for huge investment. So how do ski resorts get coverage? This motivation comes about mostly through demand of the clients, rather than the resort or telco. In many cases we see someone trying to do business in a location, and they have a connection with someone high enough up the telco management that it can be rectified.

Currently there is no real link between safety and cell phone signal, but we hope that this can change, as there is an expectation now with the public that they can get reception when they need it, even in a ski resort where it would seem obvious that you could make a phone call. One way this will be moved forward is litigation where someone was unable to get help in a location where the telco claims there is service.

Why now? Why an app?

Until the advent of tablets, smartphones were the fastest spreading technology on the planet. Smartphones have literally become part of the fabric of life in a very short space of time.

Many people have three devices: phone, tablet and laptop/computer, and we are now all connected to the internet 24 hours a day.

Australians are the most connected in the world, even more so than Americans and the British.[36]

Smartphones make up 72% of all mobile phones in Australia, and the number is rising. Of those people, 49% also own a tablet device.

Smartphone owners now think of a problem, and while deliberating the solution ask 'Is there an app for that?'

With the proliferation of technology in this way, and a natural curve towards Bring Your Own Device to work, where well above 60% of businesses[37] are supporting employees' desires to use their personal devices. Cisco's 2012[38] study shows that employees are happier and more productive when using their own devices.

This growth has prompted the phone network providers to up their game with capacity and coverage, with phones being used in some unlikely places, like the summit of Mt Kosciusko - Australia's highest peak. Ski resorts are well populated and as such make for an essential location for providers to get a signal to.

We now have many technologies available to assist with the use of devices in extreme conditions. eglove.co.uk produces a range of gloves that allow you to use a touch screen without removing your glove and getting cold hands.

Your device is kept safe through Liquipel.com who provide micro covering solutions to water proof devices, and Otterbox have hardy smash-proof and water-proof cases.

Some of the issues I considered in developing the app were:

- Patients don't know where they are, resulting in vague location descriptions
- Patients can't always send for help so have to rely on a bystander, friend or passer-by
- The time it takes for a second person to find a lift attendant or someone else to get help are critical minutes to the patient
- Direct phone numbers are helpful, but the patient may not know what they are
- People know a direct phone number is available, but forget it, and call emergency services
- Lost time means patient condition can deteriorate
- Patrollers are not recording critical information about every patient
- Patrollers are not transferring knowledge along the chain of care
- Medical teams are not always aware of the situation prior to handover, and thus cannot be prepared (remember they are usually not hospitals, so do not have equipment and staff to deal with every medical situation at hand)
- Accident investigations are mostly after the event, so photos are missing of the patient, and witness statements are changed by the passage of time
- Accident statistics are hard to compile as the information is mostly paper based

- A full understanding of incidents by resorts takes time to extract from paper
- Accuracy of locations in incident reports is broad, using interpretive descriptions, or broad area descriptions
- Understanding of demographics is underutilised
- Confirmation that preventative actions (such as fencing, signage) is confirmed only by experience and gut feel
- Lack of understanding can mean little or no reduction in accident rates
- Evidence collected in post-accident investigations are more easily proved inaccurate given the passing of time to recording
- Using the phone on the snow can be difficult
- Cell signal – mountains and heavily wooded areas (e.g. cross country skiers) can mean penetration of signal is scarce.

Other factors, including advances in technology and equipment, were incorporated into the design of the app:
- Sunlight – later smartphone models can be light adjusted and do not have glare issues
- Cold – plenty of heavy duty gloves, or glove liners have silver thread or other technology to allow use of smartphones with gloves on e.g. eGlove
- Battery life – a smartphone user knows they must keep their phone charged. Cold however reduces battery life, so a simple plastic case can help, keeping the phone near the body to keep it warm, or a full battery pack case to extend battery life all provide solutions.
- Wet – there are now solutions that allow you to fully waterproof your phone such as Liquipel. There are also cases such as the Otterbox, Lifeproof and to go even further – waterproof phones such as the Sony Experia Z and Galaxy S4 Active.

Medic52 works in three stages

1. It helps the patient/customer faster
The patient uses the existing resort smartphone application (or our standalone app) to report their location using the phone GPS, to the central dispatch system.

Optionally they can add their name, and phone number to assist in communication and patient matching on the ground.

Benefits:
- It cuts down the time for the patient to raise help
- There is no double handling of the message (via lift operators for example)
- The location of the patient is accurate using the GPS
- The patroller knows exactly where to go
- Dispatch can call the person and get more information early if necessary

2. The patroller records critical information on scene
Patrollers do not currently record much, if any, data from the scene of the incident. This means that a high (more than 50%) proportion of data is lost. In cases where more information is required for medico-legal purposes an accident investigation team must be dispatched after the event to take photos and document it.

The patroller's priority is the patient care. It may not always be practical to deal with a device as well. In these circumstances (usually life threatening or serious injury) there will always be a team of patrollers to manage the patient and scene. In good scene management the person in charge would communicate and collect information, and this is who would use the application.

In a less time critical or intensive incident (pain, sprain, knee, minor bleed, shoulder, wrist), or where extra equipment is required, the patroller can take the time to document the incident.

The minimum of information to be collected from the scene is:
- photos of scene and patient
- time and date
- GPS location

In the Medic52 application, simply creating the case deals with the GPS location, time and date, and provides a more accurate picture. Where there is no cell service the Medic52 app can function without signal for the patroller data collection. It will synchronise data as soon as it gets a signal.

The patroller is then directed to the photos screen, where they can either take a number of quick snaps, or put the phone away to concentrate on the patient.

The case can then be taken up by another patroller who arrives on scene, or at the handover after the patient is off the snow, in a more comfortable environment. This also affords the opportunity to engage the use of a tablet to enter data easily.

Future technology advances also mean that small portable devices such as a Pulse Oximeter can be connected to the device and provide interim information until a defibrillator arrives. This can be transmitted directly into the patroller's smartphone and recorded.

Benefits:
- Coverage of 100% of incidents
- Photos with the patient on scene
- Live information communicated to next level of care/manager

3) Resort & Risk Management strategy

For a resort that has to contend with the almost certain guarantee that a guest will hurt themselves, there are a lot of challenges to take in and contend with.

Keeping detailed information on every accident, and having it to hand provides benefits in many ways to a risk manager and resort management.

Current prevention strategies, more often than not, are dependant on full time staff, with many years' experience on the ground, knowing what happens and where. This pool of knowledge cannot be replaced by technology, however it does make the information hard to access, and pass on to others.

Insurance companies tend to use statistics to understand the risks involved. The more detailed the data that they can use allows for a better analysis of the risk. Many resorts currently are without detail, and thus at the mercy of the insurance market drivers (supply and demand) that will influence their pricing. This has driven many to build substantial resources to start self insuring for the smaller claims.

Benefits:
- Accessible data
- Accurate location based information
- Reportable data
- Trend identification
- Better evidence for insurance and litigation
- Development of evidence based risk management

To learn more about Medic52 visit the website
www.medic52.com

Download the app for patrollers on iTunes or Google Play

www.medic52.com/go/itunes   **www.medic52.com/go/play**

About the Author

Duncan Isaksen-Loxton is on a mission to help everyone around the world to enjoy snow sports as much as he does. His passion is to help ski patrollers do a better job through technology, make resorts safer and provide a wider benefit to the snow going public.

Born in the UK he was lucky to learn to ski in Europe with his ski instructor dad each year. He became a ski and snowboard instructor with the Professional Ski Instructors of America (PSIA) in 1998 before returning to University in the UK to study Multimedia Computing. After some travel (and more skiing) Duncan moved to Sydney, Australia in 2003.

A qualified and active Australian National Ski Patroller, and member of the *International Society for Skiing Safety*, Duncan has been working in ski area safety for 10 years. Prior to which, he spent five years working in Europe and North America as a PSIA ski and snowboard instructor. He also spent 10 summers as a volunteer lifesaver on Manly beach, one of Australia's premier beaches, as an experienced inshore rescue boat driver, instructor, and patrol captain.

Duncan is a small business owner, having built two successful mobile and web development companies in Sydney. In this professional capacity he has developed skills in understanding the problems of many industries. In close collaboration with his clients, he has created systems and solutions to solve them.

He is available for speaking and consulting, having addressed international ski industry conferences on technology and safety related topics.

Duncan would love to hear your feedback on the book, email him at **duncan@medic52.com**

Next Steps

Download and try the app, for free:

www.medic52.com/go/itunes
www.medic52.com/go/play

Communities:
Join the Medic52 Ski Safety Professionals discussion at

www.medic52.com/go/community

Other Resources:
Check out other safety and technology related resources at **www. medic52.com** and stay up to date with the development of Medic52

Consultation:
For a consultation, send an email to **duncan@medic52.com**

Notes

1. Vanat, 2013 International Report on Snow & Mountain Tourism

2. Ski Club of Great Britain http://www.skiclub.co.uk/skiclub/news/story.aspx?storyID=8972#.Uu3o9neSwq8

3. Ryan Solutions http://www.ryansolutions.com/blog/2013/value-of-season-pass-use/

4. Tracey Dickson & Tonia Gray, Risk Management in the Outdoors, 2011

5. American Association of Neurological Surgeons (www.aans.org)

6. http://www.nsaa.org/media/174893/Helmet_Fact_Sheet_10_3_2013.pdf

7. Smartrisk, 2009

8. Robert Stewart – Ski Press

9. Vanat, 2013 International Report on Snow & Mountain Tourism

10. Ibid

11. NSAA http://www.nsaa.org/press/industry-stats/industry-stats-pages/who-owns-which-mountain-resorts/

12. Rob Stewart, The Ski Press, LISTEX 2013

13. Ibid

14. Ibid

15. http://www.tpr.alberta.ca/tourism/research/docs/skiec.pdf

16. http://skiingbusiness.com/11245/newswire/press-release/rossign-ol-to-invest-13-million-in-factory-upgrades/

17. http://www.ft.com/cms/s/0/998ef8b6-ed29-11e2-8d7c-00144feab-dc0.html#axzz2rZ5Fw

18. Snowsports Industries of America http://www.snowsports.org/Re-tailers/Research/SnowSportsFactSheet

19. http://www.bcbr.com/article/20130930/NEWS/130939993

20. Dore Group http://www.thedoregroup.com/blog/u-s-ski-snowboard-resort-industry-in-2012/

21. Ibid

22. http://www.ryansolutions.com/blog/2013/vacation-streaks-vs-re-turn-rate/

23. Dore Group, 2012 http://www.thedoregroup.com/blog/u-s-ski-snowboard-resort-industry-in-2012/

24. http://www.ryansolutions.com/blog/2013/value-of-season-pass-use/

25. http://www.suva.ch/

26. http://www.deloitte.com/assets/Dcom-Switzerland/Local%20As-sets/Documents/EN/Survey/Ski/ch_en_ski_resort_survey_2011.pdf

27. Interview Dec 2012

28. Interview with Mike Campbell, Fitness Trainer & *Author of Unleash Your Alpha*, 2013

29. http://www.capitalcityweekly.com/stories/110712/out_1063359058.shtml

30. http://www.saminfo.com/blog/ski-patrollers

31. Bala Prahabar, SAP http://scn.sap.com/people/bala.prabahar/blog/2011/05/30/is-assumption-the-mother-of-all-mistakes

32. http://sls.com.au/sites/sls.com.au/files/SLS-Economic-Contribu-tion-Report-Web.pdf

33. FIS ISS Brochure 2008

34. Tracey J. Dickson, *Reduce the Injuries, Change the Culture. Insights*

from Telemarking.

35. http://blogs.esri.com/esri/arcgis/2013/07/15/smartphones-tab-lets-and-gps-accuracy/

36. Nielsen Telstra Smartphone Index http://www.news.com.au/technol-ogy/gadgets/young-aussies-spend-29-days-on-smartphones-every-year-says-new-study/story-fn6vihic-1226741142141

37. good.com State of BYOD report http://media.www1.good.com/doc-uments/Good-BYOD-Report-2013.pdf

38. https://www.cisco.com/web/about/ac79/docs/re/BYOD_Hori-zons-Global_Top10-Insights.pdf

Acknowledgements

I would like to express sincere thanks to those who helped me make this happen – those in the Key Person of Influence program and particularly Andrew Griffiths for his invaluable guidance and endless enthusiasm, and my book writing buddy Mike Campbell who has written his first book alongside me.

I am grateful to those who agreed to be interviewed: Michael Ditchburn (Click4Snow), Glenn Kirkwood (Thredbo Ski Patrol), Anthony Bradstreet (Surf Life Saving Australia), David O'Dowd (Jack Shand Chambers), Tony Paul (PicoNet Consulting) and Robin Barham (Lloyds Underwriter).

And I couldn't have put this whole thing together without the support of the many people who helped me research and make this into a real book, friends, family and passionate snow lovers, members of the Thredbo Ski Patrol and the Australian Ski Patrol Association, in particular David Varnes, Stefan Hechl, William Loxton, Dr Guy Buchanan, John Matic and David Kuhn.

Finally, a massive thank you to my wife, Heidi, for the tireless hours she puts up with talk about ski patrol and the crazy entrepreneurial spirit. You are my best friend and biggest supporter – thank you for being on this ride with me, I love you more than anything in the world.

Published in 2014 in Australia by Rasmus Pty Ltd
duncan@medic52.com
www.medic52.com

PO Box 1081
Surry Hills
NSW 2010

www.thesmartphonemedicbook.com.au

Typesetting: OpenBook Creative
Cover Design: Sweetlip Design
Editor: Nikki Cripps
Cover Photograph: Zoom in with Eden

Australia Cataloguing-in-Publication entry:

   ISBN: 9781925144116 (paperback)
   Author: Duncan Isaksen-Loxton
   Title:The Smartphone Medic (paperback)
   Subjects: Skis and skiing--Safety measures.
                Ski resorts--Safety measures.
                Skiing--Risk assessment.
                Accidents--Prevention.
                Smartphones.
                Application software.
   Dewey Number: 796.930285